Fred Manuele on Safety Management

Fred Manuele on Safety Management

A Collection from *Professional Safety*

Fred A. Manuele
Edited by Tina Angley

American Society of Safety Professionals, 520 N. Northwest Highway, Park Ridge, IL 60068
Copyright © 2018 by American Society of Safety Professionals
All rights reserved. Published 2018.

Limits of Liability/Disclaimer of Warranty
While the publisher and authors have used their best efforts in preparing this book, they make no representations or warranties with respect to the accuracy or completeness of the contents of this book, and specifically disclaim any implied warranties of merchantability or fitness for a particular purpose. The information is provided with the understanding that the authors are not hereby engaged in rendering legal or other professional services. If legal advice or other professional assistance is required, the services of a qualified professional should be sought.

Managing Editor: Rick Blanchette, ASSP
Cover and interior design: Tina Angley, ASSP

Printed in the United States of America

27 26 25 24 23 22 21 20 19 18   1 2 3 4 5 6 7 8

Library of Congress Cataloging-in-Publication Data
A catalog record for this book is available from the Library of Congress.

Contents

Editor's Preface ... ix
Foreword ... xi

Introduction ... 1

Chapter 1 ... 17
Principles for the Practice of Safety: A Basis for Discussion
Originally published July 1997, revised December 2017

Chapter 2 ... 37
Severe Injury Potential: Addressing an Often-Overlooked Safety Management Element
Originally published February 2003

Chapter 3 ... 49
Is a Major Accident About to Occur in Your Operations? Lessons to Learn From the Space Shuttle Columbia Explosion
Originally published May 2004

Chapter 4 ... 65
Risk Assessment and Hierarchies of Control: Their Growing Importance to the SH&E Profession
Originally published May 2005

Chapter 5 ... 81
The Challenge of Preventing Serious Injuries: A Proposal for SH&E Professionals
Originally published April 2006

Chapter 6   97
Prevention Through Design: Addressing Occupational Risks in the Design and Redesign Processes
Originally published October 2008

Chapter 7   123
Serious Injuries and Fatalities: A Call for a New Focus on Their Prevention
Originally published December 2008

Chapter 8   141
Acceptable Risk: Time for SH&E Professionals to Adopt the Concept
Originally published May 2010

Chapter 9   161
Accident Costs: Rethinking Ratios of Indirect to Direct Costs
Originally published January 2011

Chapter 10   177
Reviewing Heinrich: Dislodging Two Myths From the Practice of Safety
Originally published October 2011

Chapter 11   197
Management of Change: Examples From Practice
Originally published July 2012

Chapter 12   217
Preventing Serious Injuries and Fatalities: Time for a Sociotechnical Model for an Operational Risk
Originally published May 2013

Chapter 13   237
ANSI/AIHA/ASSE Z10-2012: An Overview of the Occupational Health and Safety Management Systems Standard
Originally published April 2014

Chapter 14 255
Incident Investigation: Our Methods Are Flawed
Originally published October 2014

Chapter 15 275
Culture Change Agent: The Overarching Role of OSH Professionals
Originally published December 2015

Chapter 16 289
Root-Causal Factors: Uncovering the Hows and Whys of Incidents
Originally published May 2016

Chapter 17 305
Highly Unusual: CSB's Comments Signal Long-Term Effects on the Practice of Safety
Originally published April 2017

Editor's Preface

Having worked for ASSP editing *Professional Safety* for more than 20 years, I have had many opportunities to read Fred Manuele's articles. The idea for this book began to take shape while preparing one of his articles for publication. That article, "Highly Unusual: CSB's Comments Signal Long-Term Effects on the Practice of Safety," appears as the last chapter of this book.

It struck me that the article echoed Fred's previous work. Many of the principles he discusses have been his focus for at least three decades. His emphasis has been on establishing a culture of continual improvement, macro thinking rather than micro thinking, achieving acceptable risk, performing risk assessments, prevention through design, searching out potential multiple causal factors, improving the system in which people work rather than focusing on so-called unsafe acts of employees, and the concept that reducing incident frequency does not ensure prevention of serious injuries and fatalities.

That prompted me to propose we publish a collection of Manuele's articles in one volume to highlight the significance of his work. The unanimous support from advisory committee members and staff is a testament to Fred Manuele's value to the profession. It has been an honor collaborating with him on this project.

There is no doubt that Fred is a prolific writer and respected thought leader in the safety profession whose many works have influenced some of the profession's most prominent authors, leaders, speakers and educators. His body of work has contributed significantly to the discourse of the profession, potentially influencing generations of future safety professionals.

In "Highly Unusual," Manuele calls safety professionals' attention to CSB's report on the *Deepwater Horizon* incident, saying "The writers of the report indirectly advance the state of the art in safety management." One might say the same about Fred Manuele. At the very least, this collection of articles from *Professional Safety* reflects an evolution of the safety profession. ■

Tina Angley
Editor, **Professional Safety**

Foreword

By David Walline, CSP

Tina Angley, editor of *Professional Safety*, is to be commended for deciding that the American Society of Safety Professionals should publish the compilation of select articles written by Fred Manuele that are contained in this book. Manuele writes about what matters for the safety, health and environmental professional. He has been a thought leader throughout his storied and award-winning career, encouraging safety practitioners to continue to examine and refine what they do so that eventually the practice of safety will be recognized as a profession.

Manuele has done the research necessary to explore and expound on the erroneous concepts that permeate the practice of safety. Chapters will be found in this book that are intended to eradicate those strongly held beliefs that deter the progress of our profession. He has also been a leader with respect to the principles for the practice of safety, risk assessment, prevention through design, serious injury and fatality prevention, occupational health and safety management systems and culture change to name a few. Most importantly, countless safety practitioners like myself stop and listen when Fred Manuele speaks, and his writings are read and not put down until finished.

For decades, Manuele has been a forerunner for the advancement of risk assessment into the safety management process and into the professional practice of safety. His motto has been, "It's all about risk."

If you read one new book this year on what the safety profession should believe and adopt, and why, this book is for you. If you are a safety practitioner who wants to move the state of the art forward, you may also want to share the messages in this book with others within your organization.

Each chapter could be viewed as a short course on the practice of safety that can be easily read, digested, discussed and acted on to advance safety competency and performance going forward. Some would say this compendium of articles should be required reading for every business owner, senior executive, people leader, manager, engineer, designer, incident investigator, risk assessor, human resource and safety practitioner, as well as all students in safety-related degree programs. I concur.

I believe that when one reads these compiled articles, it will be concluded that in many ways they are all interrelated. These topics should not be viewed as stand-alone subjects, but they should be collectively understood as an all-connecting process or body of work that can greatly benefit not only a dedicated safety practitioner but an entire organization that desires to advance its safety mission and performance. If a reader embraces and adopts just a few of the key practices and actions outlined in this book, I am confident that today's workers will be the beneficiaries of that to which Manuele's career has been dedicated. ■

David Walline, CSP
ASSP professional member since 1978

Introduction

BY FRED A. MANUELE, P.E., CSP

My employment and professional history, as they progressed, relate directly to the articles I wrote that were originally published in *Professional Safety*—selected by Tina Angley, editor, for inclusion in this book—but not necessarily in the order of their publication.

As World War II ended, having filled the position and entry examination requirements, I was enrolled in 1945 in a course to become a U.S. Coast Guard designated marine engineer. I continued serving as a marine engineer for about 2 years. Because of my engineering orientation, I was employed as a safety inspector in New York and, with experience and knowledge, became a safety consultant.

I was offered a position in Springfield, MA, with a company for which I eventually became engineering department superintendent. While in that role, I was qualified as a registered professional engineer by the governing board in Massachusetts with a designation in safety engineering.

To the surprise of many, the company by which we were employed in Springfield was sold. On the day the sale was announced, the purchasing company offered me a job that would take me back to New York. I decided I would return to New York only as a last resort. Travelers Insurance Co. offered me a position and I reported in the company's Hartford, CT, office for a while.

Where did they send me later? You guessed it. I was made an assistant supervisor in the engineering department of Travelers' New York office. In 1965, Travelers Insurance Co. offered me a transfer to the Chicago office as engineering manager.

Several of the insurance accounts for which I was the safety consultant were placed with Travelers by Marsh & McLennan, a large insurance broker. An offer was made to me by Marsh that I could not refuse. My career at Marsh & McLennan commenced in 1969 as manager of a group of safety and fire protection consultants in its Chicago office. Within a year, I was made an assistant vice president. A national safety and fire protection group called M&M Protection Consultants was formed in 1971 and I was appointed regional manager for the central states.

In 1976, I was appointed national manager for M&M Protection Consultants, which had grown to about 225 people. I continued in that position until I retired in 1991. My officer progression was from assistant vice president to vice president to senior vice president to managing director.

Principles

It became apparent in the latter half of the 1990s that for the practice of safety to be recognized as a profession, safety practitioners needed to establish sound theoretical and practical premises that, if applied, would be effective in:

- hazard avoidance, elimination or control;
- achieving acceptable risk levels;
- reducing injuries and illnesses.

To this day, safety practitioners have not agreed on the fundamentals for their profession or the definitions of related terms. Safety practitioners take a variety of approaches to achieving safety, each based on substantively different premises. They cannot all be right or equally effective. As John V. Grimaldi and Rollin H. Simonds wrote in their book *Safety Management*:

> Unless there is common understanding about the meaning of terms, it is clear that there cannot be a universal effort to fulfill the objective they define. (p. 10)

To promote a discussion toward establishing a sound theoretical and practical base for the practice of safety, an article titled "Principles for the Practice of Safety: A Basis for Discussion" was published in *Professional Safety* in July 1997. An updated 2017 version appears as Chapter 1 of this book. Its purpose is to encourage dialogue by those who have an interest in moving the state of the art forward. Some of the premises set forth in that article are now common usage.

Introduction

While the American National Standard for Occupational Health and Safety Management Systems (ANSI/ASSE Z10-2012) is recommended in its entirety, it is cited here because it contains a particularly basic and sound premise pertaining to the practice of safety. In the standard's planning section, Section 4.0, health and safety issues are defined as "hazards, risks and management system deficiencies." This is a seminal definition. It is basic in the practice of safety. All issues with which safety practitioners are involved relate to risks that derive from hazards that may exist because of management system deficiencies.

Progression

Studies of the causes of back injuries led me, progressively, into ergonomics, risk assessments, prevention through design, serious injury and fatality prevention, inadequacies in incident investigation, and the benefit of having a sociotechnical workplace in which the technical aspects of the work and the social (culture-climate) aspects are well balanced.

It was common practice when trying to prevent back injuries to train workers how to lift safety. But, my studies and analyses showed that training workers did not measurably reduce back injuries. Having visited locations that had reported back injuries, it became apparent that the design of the work methods often resulted in work situations in which a large percentage of the work population was overly stressed as they did what they were expected to do.

My thinking was influenced substantially by several authors. Alphonse Chapanis, who was prominent in ergonomics, coined the phrase *error provocative*. His position was that if the workplace is designed to be error provocative, you can be certain that errors will occur. An error-provocative situation is one that almost literally invites people to commit errors. Note that the premise applies to both the workplace and the work methods.

Ergonomics is design based, as is all of safety. I became much involved with and a supporter of safety practitioners becoming involved in the design of the workplace and work methods.

W. Edwards Deming, who was world renown in quality management, proposed quality achievement principles that I thought also applied to safety, such as, if you want to achieve superior quality (or safety), you must design a system in which capable people could achieve superior quality.

Other authors whose work was influential include Willie Hammer, who wrote *Handbook of System and Product Safety*, Trevor Kletz, who was the author of several notable texts, including *An Engineer's View of Human Error*, and William Johnson whose 1980 book *MORT Safety Assurance Systems* is still referenced. Dan Petersen was a colleague.

An Adherent to H.W. Heinrich's Premises

Well into the 1970s, I was a devoted advocate of certain of H.W. Heinrich's principles: that 88% of occupational incidents were caused by employee unsafe acts and if the focus was on reducing incident frequency, there would be an equivalent reduction in serious injuries. Studies that I commenced in the late 1970s showed that neither premise could be upheld.

Focusing on worker unsafe acts as the causal factor for occupational injuries and illness is still a major problem. My research indicates that for a large proportion of incident investigation reports, the principal causal factor identified is still the so-called "unsafe act" of the employee and the preventive action is usually training or retraining, or holding a group meeting to discuss and reinforce the standard operating procedure. It is acceptable in many organizations for investigations to stop at that point.

My article, "Reviewing Heinrich: Dislodging Two Myths From the Practice of Safety," was published in the October 2011 issue of *Professional Safety* and appears as Chapter 10 of this book. Response has been gratifying. So have the writings of others who support the premise that to understand incident causation the focus must shift from what the individual employee does to recognize the effect that management decisions have on the situations in which incidents occur. I give only one example of what has come to be. In *The Field Guide to Understanding Human Error*, Sydney Dekker (2006) expresses the following views on human error:

> Human error is not a cause of failure. Human error is the effect, or symptom, of deeper trouble. Human error is . . . systematically connected to features of people's tools, tasks, and operating systems. Human error is not the conclusion of an investigation. It is the starting point. (p. 15)

> If you want to understand human error, you have to dig into the system in which people work. You have to stop looking for people's shortcomings. (p. 17)

I continue to make this point: If the environment that has been constructed, which derives from the design of the physical aspects of the workplace and the design of work methods, and operational and management influences, results in logical, normal behavior that is considered unsafe, a causation model should give prominent emphasis to the decision making out of which those environmental causal factors arose.

Risk Assessments

It became apparent in the 1970s that decision makers could be influenced into looking into the design of their workplaces and work methods only if convincing data were developed and presented to them. Thus, risk assessments would have to be made and that idea was introduced to safety practitioners through articles and books.

But, then came OSHA and the behaviorists. For much of the 1970s and through the 1990s, complying with OSHA requirements dominated the activity of many safety practitioners. Behavior-based safety had its prominence through the 1990s. Adherents to behavior-based safety, of which there were many, had a significant influence on the safety profession and were an obstacle to promoting the significance of the design of work systems and work methods and risk assessments.

But, events of the recent past indicate that several of the big hitters in behavior-based safety have revised their positions, and now talk and write about taking a systems approach to safety management.

At an ASSE conference in the early 1990s, I met Michael Taubitz, who was then at General Motors. During our conversation, Michael said he had written to the then executives at NSC encouraging that the organization undertake an activity related to a system then in place at GM called design-in-safety. This activity focused largely on task-based risk assessment and involved all personnel levels. Based on our discussion, I agreed to approach NSC executives and attempt to convince them to do something comparable to what Michael had proposed.

Decision makers at NSC gave me authority to form a committee to study the feasibility of promoting the concept of addressing safety and environmental matters in the design process. In 1995, NSC established The Institute for Safety Through Design. Hazard identification, hazard analysis and risk assessments are at the core of safety through design. For the institute, its definition of safety through design was:

> The integration of hazard analysis and risk assessment methods early in the design and engineering stages and taking the actions necessary so that risks of injury or damage are at an acceptable level.

Much was accomplished by the institute. Seminars, workshops and symposia were held. Proceedings were published. Presentations were delivered at safety conferences. *Safety Through Design* was published by NSC, coedited by Wayne Christensen and me. In accordance with its sunset provisions, the institute was disbanded in 2005.

Slowly, the shift in the practice of safety is toward risk-based decision making. Progress made in the past 25 years with respect to safety practitioners doing risk assessments is a pleasure to behold. Through the early 1990s, it would be the exception for an article to be included in the relative safety literature that focused on risk or risk acceptance. Today, such articles frequently appear. A demonstrative example: In the August 2016 issue of *Professional Safety*, Tom Cecich's President's Message was titled "It's All About Risk."

My article, "Risk Assessment and Hierarchies of Control: Their Growing Importance to the SH&E Profession," was published in the May 2005 issue of *Professional Safety* and appears as Chapter 4 of this book.

In 2013, ASSE decision makers recognized the emerging importance of risk assessments, creating the Risk Assessment Institute. Every safety practitioner should be aware that risk assessments are potential problem identification techniques.

Prevention Through Design

In 2006, several participants in the activities of The Institute for Safety Through Design and others received an e-mail from a National Institute for Occupational Safety and Health (NIOSH) official encouraging participation in an initiative for prevention through design. In 2008, NIOSH announced a major initiative to "develop and approve a broad, generic voluntary consensus standard on prevention through design that is aligned with international design activities and practice."

I volunteered to lead that endeavor and obtained support from ASSE's Standards Development Committee. On Sept. 1, 2011, American National Standards Institute (ANSI) approved the standard ANSI/ASSE Z590.3-2011, Prevention through Design: Guidelines for Addressing Occupational Hazards and Risks in the Design and Redesign Processes (reaffirmed 2016).

There is now much greater acceptance of the idea that the most effective, economical way to address hazards and risks is doing so early in the design and redesign stages and taking the actions necessary so that risks of injury, illness or damage are at an acceptable level. Several examples follow:

•At a session during ASSE's annual conference, it was a pleasant, rewarding surprise to note the large number of people who raised their hands when asked how many were involved as consultants in the design process.

•For an ASSE-sponsored training program on prevention through design, the number of people who wanted to attend was more than could be accommodated.

•In December 2017, Dodge Data and Analysis issued a report titled "Safety Management in the Construction Industry 2017." Prevention through design is a major part of the report.

It is important to note that activities at NIOSH are limited to occupational safety and health. But, by intent, the terminology in Z590.3 was kept broad enough so that the guidelines could be applicable to all hazards-based needs, including product safety, environmental controls and property damage that could result in business interruption.

I addressed the subject in the article, "Prevention Through Design: Addressing Occupational Risk in the Design and Redesign Processes." It was published in the October 2008 issue of *Professional Safety* and appears as Chapter 6 of this book.

Serious Injury, Illness and Fatality Prevention

One of Heinrich's premises was that focusing on reducing incident frequency would result in an equivalent reduction in serious injuries. That premise is deeply embedded in the minds of many safety practitioners. In the late 1980s and early 1990s, through analysis of incident investigation reports and data published by Bureau of Labor Statistics and National Council on Compensation Insurance, some safety practitioners recognized that while incident frequency was coming down, the reduction largely involved less serious injuries and illnesses.

Encouraging safety practitioners to recognize that serious injury, illness and fatality prevention required specifically tailored potential identification and action measures became a major undertaking for me.

My article titled, "Severe Injury Potential: Addressing an Often-Overlooked Safety Management Element," was published in the February 2003 issue of *Professional Safety* and appears as Chapter 2 of this book. I continued writing about the need to introduce methods to identify serious injury, illness and fatality potential. My theme was that data in support of such activities would be developed principally through hazards identification, hazard analyses and risk assessments.

Having recognized that lessons could be learned from *Columbia* space shuttle disaster, in a May 2004 *Professional Safety* article, I asked safety practitioners "Is a Major Accident About to Occur in Your Operations? Lessons to Learn From the Space Shuttle *Columbia* Explosion." It appears as Chapter 3 of this book.

That was followed by the articles, "The Challenge of Preventing Serious Injuries: A Proposal for SH&E Professionals," in April 2006 (Chapter 5) and "Serious Injuries and Fatalities: A Call for a New Focus on Their Prevention," in December 2008 (Chapter 7).

My latest article on this subject, published in the May 2013 issue of *Professional Safety*, is "Preventing Serious Injuries and Fatalities: Time for a Sociotechnical Model for an Operational Risk Management System" (Chapter 12). Gradually, recognition is evolving that serious injury, illness and fatality potential needs special attention. That recognition arises in part from an awareness that has developed in many companies that OSHA-type statistics have plateaued, and that having such stellar rates does not ensure that adequate controls are in place to prevent serious injuries, illnesses and fatalities.

For more than 40 years, an overemphasis has been placed on achieving low OSHA incident rates without recognizing severity potential. As an example, companies that are members of trade associations may be in competition with each other based on OSHA-type rates.

When achieving low OSHA incidence rates is deeply rooted within a company's culture and doing so is considered sufficient for controlling all types of incidents, dislodging that concept will require long-term effort. A culture change is not a one-time program; it is a lengthy journey that must engage all employment levels in an organization.

It was a pleasure to note that the agenda for a January 2017 planning meeting to consider the next version of Z10 included a suggestion by a participant that guidelines be included with respect to preventing serious injuries and fatalities.

Macro Thinking: A Sociotechnical Workplace

Macro thinking and the desirability of having a balanced sociotechnical workplace were introduced in "Preventing Serious Injuries and Fatalities: Time for a Sociotechnical Model for an Operational Risk Management System" (Chapter 12). My interest in the history of the premise that an organization that achieves a balance between its social aspects (culture, climate) and its technical aspects would have a less risky (and more efficient) operation arose out of the realization that causal factors (plural) could derive from both what people do or do not do and the technical aspects of operations.

It also became apparent that macro thinking would be necessary for the identification of the reality of incident causal factors. My composite definition of macro thinking applied to a sociotechnical system follows.

> Applied macro thinking takes a holistic approach to analysis that focuses on the whole and its parts at the same time and the way a system's parts interrelate. Macro thinking contrasts with an analytical process that addresses a technical or social aspect of a system separately—micro thinking—without considering the relationship of that aspect to the system as a whole.

My observation was that in an organization that applied a sociotechnical systems approach, a macro thinking approach, it would be understood that:

- technical and social systems in an organization are inseparable parts of a whole;
- parts are interrelated and integrated;
- changes made in one system may affect others;
- the organization and the people who work in it are not well-served if, when resolving a risk situation, the subject is considered in isolation rather than as part of an overall system.

Emphasis is on the importance of the whole and the interdependence of its parts. In an effective sociotechnical system, management recognizes the interdependence of the technical and social aspects of operations and integrates them, and understands that macro thinking would have to be applied in studying an incident to identify the totality of its causal factors. A feedback process would be created to monitor alignment.

Micro thinking focuses on the direct or immediate contributing factors in a hazard/risk situation and does not explore shortcomings in the management systems connected to them. Incident investigation reports that stop with the identification of unsafe acts of employees as causal factors are examples of a narrow and micro thought process.

Acceptable Risk

Although the term *acceptable risk* is used in many jurisdictions throughout the world, research indicated that many safety practitioners shy away from the use of the term. People engaged in the practice of safety must understand the concept as a goal to be reached by decision makers. They should have a guideline that serves as a base in answer to the question, when have we gone far enough?

Achieving agreement on the definition of acceptable risk is necessary for the practice of safety to be recognized as a profession. My article on the subject is titled, "Acceptable Risk: Time for SH&E Professionals to Adopt the Concept." It was published in May 2013 and appears as Chapter 8 of this book. In the past several years, fortunately, the term acceptable risk appears much more frequently in safety-related literature.

Accident Costs

Another myth that permeates the practice of safety is also based on H.W. Heinrich's writings. In a 1931 edition of his book, *Industrial Accident Prevention*, and in later editions, he said that the indirect costs of accidents are at a ratio of 4 to 1 in relation to direct costs.

No studies support Heinrich's ratios. Also, no ratios on direct and indirect costs published prior to 1995 are currently valid because the increase in direct costs—indemnity and medical costs—has substantially exceeded the increase in indirect costs.

Presentations to management on the costs of worker injuries and illnesses can be attention-getting and convincing, provided the data are plausible and can be supported with suitable references. It is unprofessional to use data that are knowingly unsubstantiated in a report.

In his book, *Techniques of Safety Management*, Dan Petersen (1989) soundly expressed concern over the use of an indirect to direct cost ratio for which supporting data are questionable.

> Although hidden costs are very real, they are very difficult to demonstrate. To say arbitrarily to management that they amount to four times the insurable costs is asking for trouble. If management asks for proof, you can only say, "Heinrich said so." Management wants facts—not fantasy. Without proof, hidden costs become fantasy. (p. 132)

Through extensive research, only one credible study was found. An article published in January 2011 titled, "Accident Costs: Rethinking Ratios of Indirect to Direct Costs," Chapter 9 of this book, uses this study as a base to establish fairly current ratios.

I hoped that readers would take a more professional approach as they cite ratios. Only estimates could be developed and those would have changed in the past 7 years, the result being that indirect costs of occupational injuries and illnesses are likely notably less that the direct costs.

Management of Change

Reviews of incident investigation reports, mostly for serious injuries and illnesses, support the need for and the benefit of having management of change (MOC) systems. They show that a significantly large share of incidents resulting in serious injury and illness occurs:
- when unusual and nonroutine work is being performed;
- in nonproduction activities;
- in at-plant modification or construction operations (e.g., replacing a motor weighing 800 lb to be installed on a platform 15 ft above the floor)
- during shutdowns for repair and maintenance and startups;
- where sources of high energy are present (e.g., electrical, steam, pneumatic, chemical);
- where upsets occur: situations going from normal to abnormal.

Having an effective MOC system in place would have served well to reduce the probability of serious injuries and fatalities occurring in these operational categories.

Other than for the chemical and petroleum industries, there are no national standards requiring that MOC systems be in place.

A collection was made of MOC systems that had been installed in other than chemical and petroleum operations. A July 2012 article titled, "Management of Change: Examples From Practice," presents a summary of the advantages of having MOC systems within an operations risk management system and gives several real world and practical examples of such systems, from very simple to very involved. It appears as Chapter 11 of this book.

ANSI/AIHA/ASSE Z10-2012

It is possible that the writers of the Z10, Occupational Health and Safety Management Systems, standard may not have realized the favorable impact their work would have on the practice of safety. The standard is broadly accepted and applied.

My article about the many changes in the second edition of the Z10 standard was published in the April 2014 issue of *Professional Safety*. "ANSI/AIHA/ASSE Z10-2012: An Overview of the Occupational Health and Safety Management Systems Standard" appears as Chapter 13 of this book. One of my books, *Advanced Safety Management: Focusing on Z10 and Serious Injury Prevention*, principally because of its extensive coverage of Z10, is used in several university safety degree programs.

Every safety practitioner should have a copy of the Z10 standard and be familiar with its content. The following theme is prominent throughout all sections of Z10: Processes for continual improvement are to be in place and implemented to ensure that:

•hazards are identified and evaluated;

•risks are assessed and prioritized;

•management system deficiencies and opportunities for improvement are identified;

•risk elimination, reduction or control measures are taken to ensure that acceptable risk levels are attained.

That is sound thinking. As noted, the definition of occupational safety and health issues in Z10—subjects that are at the basis of the practice of safety—are hazards, risks and management system deficiencies.

Incident Investigation

Beginning in the 1970s, I undertook several studies of the quality of incident investigation reports and had to conclude that even in some of the largest companies, their worth was less than stellar. Causal factors identified in a large percentage of the cases focused on the so-called worker unsafe act rather than on the system in which the individual works.

In the organizations that participated in my studies, mostly Fortune 500 companies (and likely in most other entities), opportunities were missed for risk reduction and incident avoidance because of the shallowness of the investigations. Often, the preventive actions recommended and taken were misdirected and the reality of the causal factors continued to exist.

I continue to implore safety practitioners to evaluate and modify the causation models they have chosen and make the necessary adjustments to emphasize the issues, which are defined in Z10's planning section as hazards, risks and deficiencies in management systems. Safety practitioners should be willing to ask, if incident investigation is not done well and the true causal factors are not identified, what else do they do that is not soundly based?

My article, "Incident Investigation: Our Methods Are Flawed," was published in the October 2014 issue of *Professional Safety*. I was pleased with the reception it received in several countries. At least some safety practitioners were willing to do the self-examination I proposed.

In that article, which appears as Chapter 14 in this book, I recommended that the five-why system for problem solving be used because it is easy to learn and apply, and I continue to strongly recommend its adoption.

I was impressed by the reality of a statement in Dekker's (2006) *The Field Guide to Understanding Human Error*, one that must be considered by safety practitioners who give advice when incident investigations are made.

> Where you look for causes depends on how you believe accidents happen. Whether you know it or not, you apply an accident model to your analysis and understanding of failure. An accident model is a mutually agreed, and often unspoken, understanding of how accidents occur. (p. 81)

How safety practitioners look for causal and contributing factors is a reflection of their adopted causation model, and they do have such a model or models, whether recognized or unrecognized. Their models relate to what they have learned and their beliefs with respect to how incidents happen.

Safety Practitioners as Culture Change Agents

My recognition that the overarching role of a safety practitioner is that of a culture change agent resulted from conversations with a professor who was teaching classes using my books. Since I wrote an article on the subject, it can be concluded that she was successful.

That article, published in December 2015 and titled "Culture Change Agent: The Overarching Role of OSH Professionals," makes the case that every proposal made by a safety professional for improvement in an occupational safety and health management system pertains to a deficiency in a system or process. The deficiency can be corrected only if there is a modification in an organization's culture—a modification in the way things get done, a modification in the *system of expected performance*. Thus, the primary role for a safety professional is that of a culture change agent. What does the term *overarching* mean? A composite definition, as found in dictionaries, is:

> Encompassing everything; embracing all else; including or influencing every part of something.

This premise that the overarching role of a safety professional is that of a culture change agent applies universally to all who give advice on improving safety management systems. There are no exceptions. Definitions of a change agent are numerous. This definition is a composite that fits well with the safety professional's position.

> A change agent is a person who serves as a catalyst to bring about organizational change. Change agents assess the present, are controllably dissatisfied with it, contemplate a future that should be, and act to achieve the culture changes necessary to achieve the desired future.

The article, which appears as Chapter 15 of this book, is conceptual. It pertains to how safety practitioners perceive their roles and what it takes to achieve a culture change, which is not easy to do.

Root-Causal Factors

Research for the May 2016 article, "Root-Causal Factors: Uncovering the Hows and Whys of Incidents," was undertaken because of thought-provoking comments made by Erik Hollnagel and Sidney Dekker concerning accident causation. For the article, reference is made to two of their books. Hollnagel is the author of *Barriers and Accident Prevention*; Dekker wrote *The Field Guide to Understanding Human Error*. Both authors have attained stature and their writings should be considered seriously by safety professionals. Those authors say that:

1) There are several ways in which an accident can be described and understood, and the cause-effect assumption is perhaps the least attractive option (Hollnagel, 2004, p. 26).

2) The tendency to look for causes rather than explanations is often reinforced by the methods that are used for accident analysis. The most obvious example is the principle of root-cause analysis (Hollnagel, 2004, p. 26).

3) Root cause is a meaningless concept (Hollnagel, 2004, p. 28).

4) There is no "root" cause (Dekker, 2006, p. 77).

5) What you call "root cause" is simply the place where you stop looking further (Dekker, 2006, p. 77).

6) Where you look for causes depends on how you believe accidents happen. Whether you know it or not, you apply an accident model to your analysis and understanding of failure (Dekker, 2006, p. 81).

My article, which appears as Chapter 16 of this book, reviews and comments on statements made by the authors cited and presents a concept with respect to root-causal factor determination that can be practicably applied.

Identifying causal/contributing factors for incidents has been a basic element in safety management systems since initiatives were first undertaken to reduce occupational injuries. Simply stated, the purpose of an incident investigation is to learn from history and to make improvements to overcome the management system deficiencies noted in investigation reports.

As I have done elsewhere, a strong emphasis is given to the practicality of using the five-why problem-solving technique when making incident investigations.

Highly Unusual

Several comments pertaining to operations risk management are made by the U.S. Chemical Safety and Hazard Investigation Board (CSB) in the agency's April 12, 2016, final report on an explosion and fire that occurred at the Macondo *Deepwater Horizon* rig in the Gulf of Mexico on April 20, 2010. That incident resulted in 11 fatalities, 17 injuries and extensive environmental damage.

Those comments were considered to be highly unusual for a governmental agency to make. It was thought that they could possibly be harbingers indicating that, over time, entities will want to revise their accountability levels and the content of their operations risk management systems that are to protect people, property and the environment. Those possible long-term indicators are:

•Boards of directors and senior managements will be held more accountable when major incidents occur.

•Companies will be expected to have effective, realistic and continual risk assessment and risk reduction processes in place.

•Residual risks are to be acceptable and meet ALARP criteria. (ALARP is defined in the prevention through design standard Z590.3 as the level of risk that can be further lowered by an increase in resource expenditure that is disproportionate in relation to the resulting decrease in risk.)

It was the intent of my article, "Highly Unusual: CSB's Comments Signal Long-Term Effects on the Practice of Safety," published in April 2017, to bring the CSB statements to the attention of safety practitioners. The article appears as Chapter 17 of this book.

I desire very much to have the practice of safety recognized as a profession and that theme is foundational for all of the articles selected for this book. It is my hope that safety professionals find guidance in this collection of articles to further that cause. ■

Fred A. Manuele, P.E., CSP
President, Hazards Limited

References

Grimaldi, J.V. & Simonds, R.H. (1989). *Safety management* (5th ed.). Homewood, IL: Irwin Press.

Petersen, D. (1989). *Techniques of safety management* (3rd ed.). Goshen, NY: Aloray Inc.

Dekker, S. (2006). *The field guide to understanding human error.* Burlington, VT: Ashgate Publishing Co.

Hollnagel, E. (2004). *Barriers and accident prevention.* Burlington, VT: Ashgate Publishing.

1 Principles for the Practice of Safety
A Basis for Discussion

Originally published July 1997, revised December 2017

For the practice of safety to be recognized as a profession, it must have a sound theoretical and practical base, the application of which will be effective in avoiding, eliminating or controlling hazards and, thereby, achieving acceptable risk levels. This author believes that there is a generic base for the work of safety practitioners that must be understood and applied if they are to give management the appropriate advice. But safety practitioners have not yet agreed on those fundamentals or on the definitions of related terms. As Grimaldi and Simonds (1989) wrote in *Safety Management*:

> Unless there is common understanding about the meaning of terms, it is clear that there cannot be a universal effort to fulfill the objective they define. (p. 10)

Safety practitioners take a variety of approaches to achieving safety, each based on substantively different premises, which cannot all be right or equally effective. To promote a discussion toward establishing a sound theoretical and practical base for the practice of safety, presented here is a list of general principles, statements and definitions that is believed to be sound. The list is a beginning; it is not complete.

It is hoped that this beginning will encourage dialogue by those who have an interest in moving the state of the art forward.

Issues: Generally Applicable

While ANSI/AIHA/ASSE Z10-2012, Occupational Health and Safety Management Systems, is recommended in its entirety, it is cited here because it contains a particularly basic and sound premise pertaining to safety-related issues. In Section 4.0 of the standard on planning, safety and health issues are defined as "hazards, risks and management system deficiencies" (p. 9). That is a seminal definition. It is basic in the practice of safety. All issues with which safety practitioners are involved relate to risks that derive from hazards that may exist because of management system deficiencies.

On Hazards

1) Hazards are the generic base of, the justification for the existence of, the entirety of the practice of safety. If there are no hazards, safety professionals need not exist.

2) The entirety of purpose of those responsible for safety, regardless of title, is to manage their endeavors with respect to hazards so that the risks deriving from those hazards are acceptable.

3) A hazard is defined as the potential source of harm.

4) Hazards include the characteristics of things (e.g., equipment, technology, processes, dusts, fibers, gases, materials, chemicals) and the actions or inactions of persons that have the potential to harm or damage people, property or the environment.

5) By definition, all risk controversies concern the risks associated with some hazard. The term hazard is used to describe any activity or technology that produces risk (Fischhoff, 1989).

6) Two considerations are necessary in determining whether a hazard exists: 1) Do the characteristics of the things or the actions or inactions of people present the potential for harm or damage? 2) Can people, property or the environment be harmed or damaged if the potential is realized?

7) Every activity undertaken as a part of an operational risk management system should serve to avoid, eliminate or control hazards so that the risks deriving from those hazards are acceptable.

8) Hazard analysis is the most important safety process in that if it fails all other processes are likely to be ineffective (Johnson, 1980).

9) In the hazard analysis process, one must assess the exposure. An exposure assessment would determine the number of people who could be affected and how often, as well as the extent of the probable property damage and environment damage.

10) If hazard identification and analysis do not relate to actual causal factors, corrective actions will be misdirected and ineffective.

11) If a hazard is not avoided, eliminated or controlled, its potential may be realized and a hazards-related incident or exposure may occur that will likely result in harm or damage, depending on exposures.

12) Hazards and risks are most effectively and economically avoided, eliminated or controlled in the design and redesign processes.

13) A hazard-related incident is an unplanned, unexpected process of multiple and interacting events, deriving from the realization of uncontrolled hazards and occurring in sequence or in parallel that is likely to result in harm or damage.

14) Hazards-related incidents or exposures, even the ordinary and frequent, may have multiple and interacting causal factors.

Defining Risk, Acceptable Risk and Safety

1) Risk is defined as an estimate of the probability of a hazard-related incident or exposure occurring and the severity of harm or damage that could result.

2) Probability is defined as an estimate of the likelihood of an incident or exposure occurring that could result in harm or damage for the selected unit of time, events, population, items or activity being considered.

3) Severity is defined as an estimate of the magnitude of harm or damage that could reasonably result from a hazard-related incident or exposure. (Severity considerations include injury and illness to people, damage to property and the environment, business downtime and loss of business.)

4) Acceptable risk is that risk for which the probability of an incident or exposure occurring and the severity of harm or damage that may result are as low as reasonably practicable (ALARP) in the setting being considered.

5) ALARP is that level of risk that can be further lowered only by an expenditure that is disproportionate in relation to the resulting decrease in risk.

6) Safety is defined as freedom from unacceptable risk. (As used in this article, the term *safety* applies to all aspects of occupational risks, to injuries to nonemployees, to the possibility of property damage and business interruption, and to environmental safety.)

7) All risks to which the practice of safety applies derive from hazards; there are no exceptions.

8) It is impossible to attain a risk-free environment. Even in the most desirable situations, residual risk will remain after application of the best practical prevention methods.

9) Setting a goal to achieve a zero-risk environment may seem laudable but doing so requires chasing a myth.

10) Residual risk is the risk that remains after risk reduction measures have been implemented.

11) The professional practice of safety requires consideration of two distinct aspects of risk: avoiding, eliminating or reducing the probability of the occurrence of a hazard-related incident or exposure; and reducing the severity of harm or damage if an incident or exposure occurs.

12) For an operation to proceed, its risks must be determined acceptable.

Risk Assessment

1) Risk assessment is a process beginning with hazard identification and analysis that produces an estimate of the severity of harm or damage that may result if an incident or exposure occurs, followed by an estimate of the probability of an incident or exposure occurring, and concluding with a risk category (e.g., low, moderate, serious, high).

2) Risk assessment is the cornerstone of initiatives to prevent harm or damage to people, property and the environment. In August 2008, the EU gave importance to risk assessment through a bulletin that is now available in "Prevention and Control Strategies," issued by OSHWiki, which says:

> Risk assessment is the cornerstone of the European approach to prevent occupational accidents and ill health. It is the start of the health and safety management approach. If it is not done well or not at all the appropriate preventative measures are unlikely to be identified or put in place.

3) If risk assessments, the first step in operational risk management, are not done well or not done at all, the appropriate preventive measures are unlikely to be identified and put in place.

4) In producing the measure that becomes a statement of risk, the assessor must determine the existence of one or more hazards; exposure to the hazard; frequency of endangerment of that which is exposed to the hazard; severity of the consequences should the hazard's potential be realized (the extent of harm or damage to people, property or the environment); and probability of the hazard being realized.

5) A successful communication with management on risk is not possible until an understanding has been reached on the meaning of the term as it is to be used in those communications.

6) Risk assessments can be considered the first step in problem solving in that they can be used to identify potential precursors of incidents that could result in serious injuries and fatalities before such incidents occur.

7) Although a safety practitioner may present logically developed data on risks, risk reduction decisions might not be made on that data alone, particularly when dealing with perceptions of risk that employees or members of the public have.

8) When risk decisions are made, it is not unusual for the decision process to be influenced by elements of fear, dread and the perceived risks of employees, management personnel, the community and the public.

9) Risk assessments would be made for all aspects of operational risk management: occupational safety, occupational health, damage to property or the environment, product safety, all aspects of transportation safety, safety of the public, health physics, system safety, fire protection engineering and business interruption avoidance.

Defining the Practice of Safety

1) For the practice of safety to be recognized as a profession, it must serve a declared and understood societal purpose, and clearly establish what the outcome of applying the practice should be.

2) The practice of safety:

•serves the societal need to prevent or mitigate harm or damage to people, property, and the environment;

•requires knowledge and skill as each of the following pertain to occupational safety and health, property damage and environmental concerns: applied engineering; applied sciences; management principles; information and communications; and legal and regulatory affairs;

•is accomplished through a) anticipating, identifying and evaluating hazards and assessing the risks that derive from them; and b) taking actions to avoid, eliminate or control those hazards;

•has as its ultimate purpose attaining a state for which the risks are judged to be acceptable.

3) Whatever the field of a hazard-related endeavor or the name given to it, the cited definition is applicable. That includes occupational safety, occupational health, environmental affairs, product safety, all aspects of transportation safety, safety of the public, health physics, system safety and fire protection engineering.

4) While safety practitioners may undertake many tasks in their work, the purpose of each is to have the attendant risks be acceptable.

5) The four major stages in operational risk management to which this definition of the practice of safety applies are:

•Preoperational stage, in the initial planning, design, specification, prototyping and construction processes, where the opportunities are greatest, and the costs are lowest for hazard and risk avoidance, elimination, reduction or control.

•Operational stage, where hazards and risks are identified and evaluated, and mitigation actions are taken through redesign initiatives or changes in work methods before incidents or exposures occur.

•Post-incident stage, where organizations investigate incidents and exposures to determine the causal factors, which will lead to appropriate interventions and acceptable risk levels.

•Post-operational stage, when an organization undertakes demolition, decommissioning or reusing/rebuilding operations.

Hierarchy of Controls

1) A hierarchy of controls provides a systematic way of thinking, considering steps in a ranked, sequential order, to choose the most effective means of avoiding, eliminating or reducing hazards and their associated risks.

2) Acknowledging the premise that risk reduction measures should be considered and taken in a prescribed order represents an important step in the evolution of the practice of safety.

3) In all four stages of operational risk management, a hierarchy of controls is to be applied to achieve acceptable risk levels.

4) With respect to the seven levels of the hierarchy of controls (Figure 1), the first through fourth are most effective because they are preventive actions that eliminate or reduce risk by design, elimination, substitution and engineering measures; rely the least on human behavior or the performance of personnel; and are less likely to be defeated by managers, supervisors or workers.

5) Actions in the fifth, sixth and seventh levels are contingent actions that rely greatly on the performance of personnel for their effectiveness and, thereby, are less reliable.

On Achieving the Theoretical Ideal for Safety

1) The theoretical ideal for safety is achieved when all risks deriving from hazards are at an acceptable level.

Chapter 1: Principles for the Practice of Safety

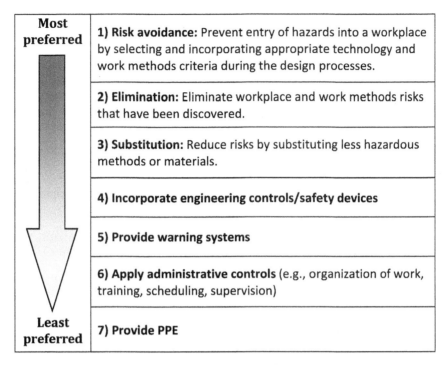

Figure 1: Risk Reduction Hierarchy of Controls
Adapted from "Prevention Through Design: Guidelines for Addressing Occupational Hazards and Risks in Design and Redesign Processes [Z590.3-2011(R2016]," by ANSI/ASSE, 2016, Park Ridge, IL: Author.

2) That definition serves, generally, as a mission statement for the work of safety professionals and as a reference against which each of the many activities in which they engage can be measured.

3) With minimal modification, a statement in *Why TQM Fails and What to Do About It* provides a basis for review to determine how near operations are to achieving the theoretical ideal for safety. In the following quotation, the word safety appears twice. In the first instance, it replaces TQM; in the second, it replaces quality:

> When safety is seamlessly integrated into the way an organization operates on a daily basis, safety becomes not a separate activity for committees and teams, but the way every employee performs job responsibilities. (Brown, Hitchcock & Willard, 1994, p. 79)

4) When safety is seamlessly integrated into the way an organization functions on a daily basis, a separately identified safety management system is not needed, theoretically, since all actions required to achieve safety would be blended into operations.

5) Thus, the theoretical ideal for a safety management system is nothing.

On Organizational Culture

1) Management creates the culture for operational risk management, whether positive or negative.

2) An organization's culture, translated into an expected performance system, determines management's commitment, or lack of commitment, to safety and the level of safety to be attained.

3) What the board of directors or senior management decides is acceptable for the prevention and control of hazards reflects its culture.

4) Principal evidence of an organization's culture with respect to operational risk management is demonstrated through the design decisions for the facilities, hardware, equipment tooling, materials, processes, configuration and layout, work environment and work methods.

5) An organization's culture consists of its values, beliefs, legends, rituals, mission, goals, performance measures and sense of its responsibility to its employees, customers and community, all of which are translated into a system of expected performance.

6) Management obtains, as a derivation of its culture (an extension of its expected performance system), the hazards-related incident experience that it establishes as acceptable. For employees, acceptable is their interpretation of what management does.

7) An organization's culture, translated into an expected performance system, determines management's level of involvement; accountability system; provision or nonprovision of the necessary resources; safety policy; safety organization; standards for workplace and work methods design; requirements for continuous improvement; and the climate that prevails concerning management and personnel factors (e.g., leadership, training, communication, adherence to safe work practices).

8) Management commitment is questionable if the accountability system does not include safety performance measures that affect the well-being of those responsible for results or if resources are not adequate to maintain acceptable risk levels throughout the organization.

9) What management does, rather than what it says, defines the actuality of its commitment or noncommitment to safety.

10) Principal evidence of an organization's culture with respect to safety is demonstrated through the design decisions that determine what the facilities, hardware, equipment, tooling, materials, configuration and layout, work environment and work methods are to be.

11) If the design of a system (facilities, equipment, work methods) does not achieve acceptable risk levels, it is unlikely that system can attain superior results.

12) Where the culture demands superior safety performance, the sociotechnical aspects of operations are well balanced (design and engineering, management and operations, and the task performance aspects).

13) An organization will only achieve major improvements in safety if a culture change takes place, only if major changes occur in the reality of the performance system.

Concerning Leadership, Training and Persuasion

1) Effective leadership, training, communication, persuasion and discipline are vital aspects of safety management, without which an organization cannot achieve superior results.

2) However, training and persuasion are often erroneously applied as solutions to problems, with unrealistic expectations. Such personnel actions have limited effectiveness when causal factors derive from workplace and work methods design decisions. (It is recognized that, in certain situations, personnel actions are the only preventive actions an employer can take.)

3) In *Training in the Workplace: Strategies for Improved Safety and Performance*, Heath and Ferry (1990) observe:

> Employers should not look to training as the primary method for preventing workplace incidents that result in death, injury, illness, property damage or other downgrading incidents. They should see if engineering revisions can eliminate the physical safety and health hazards entirely. (p. 6)

4) As an idea, the substance, but not the precise numbers, of what is called Deming's 85-15 rule, applies to all aspects of the practice of safety. In *The Deming Management Method*, Walton (1986) says:

> Deming's 85-15 Rule holds that 85% of the problems in any operation are within the system and are the responsibility of management, while only 15% lie with the worker. (p. 242)

5) In *Out of the Crisis*, this is how Deming (1986) treats the subject:

> I should estimate in my experience most troubles and most possibilities for improvement add up to proportions something like this: 94% belong to the system (responsibility of management); 6% special. (p. 315)

6) The premise is valid that many problems in any operation are systemic, deriving from the workplace and the work methods created by management, and can be resolved only by management. Responsibility for only the relatively small remainder lies with the worker.

7) Extrapolating from Deming, many causal factors for hazards-related incidents are systemic and a small minority will be principally employee focused.

8) System problems can only be corrected by a redesign of that system, and management is responsible for it. If system design and work methods design are the problem, then employees can help primarily by identifying problems.

9) In *Out of the Crisis*, Deming (1986), referencing Juran, speaks of workers being "handicapped by the system":

> The supposition is prevalent the world over that there would be no problems in production or in service if only our production workers would do their jobs in the way that they were taught. Pleasant dreams. The workers are handicapped by the system, and the system belongs to management. It was Dr. Joseph M. Juran who pointed out long ago that most of the possibilities for improvement lie in action on the system and that contributions of production workers are severely limited. (p. 134)

10) While employees should be trained and empowered up to their capabilities and encouraged to make contributions to safety, they should not be expected to do what they cannot.

11) While safety is a line responsibility, operating level achievements by management are limited by previously made workplace and work methods design decisions.

12) If the design of the system presents excessive operational risks for which the cost of retrofitting is prohibitive, administrative controls (perhaps the only actions that can be taken) will achieve less-than-superior results.

Human Errors, Unsafe Acts: Revised Views

Since the original article, titled, "Principles for the Practice of Safety," was published in 1997, several individuals have made substantial revisions with respect to the prevention of human errors (unsafe acts). Safety professionals should carefully consider those revisions and the thought process that led to them.

1) Dekker's (2006) comments on human error, as found in *The Field Guide to Understanding Human Error*, are pertinent.

> Human error is not a cause of failure. Human error is the effect, or symptom, of deeper trouble. Human error is systematically connected to features of people's tools, tasks and operating systems. Human error is not the conclusion of an investigation. It is the starting point. (p. 15)

> Sources of error are structural, not personal. If you want to understand human error, you have to dig into the system in which people work. (p. 17)

> Error has its roots in the system surrounding it, connecting systematically to mechanical, programmed, paper-based, procedural, organizational and other aspects to such an extent that the contributions from system and human error begin to blur. (p. 74)

The view that accidents really are the result of long-standing deficiencies that finally get activated has turned people's attention to upstream factors, away from frontline operator "errors." The aim is to find out how those "errors," too, are a systematic product of managerial actions and organizational conditions. (p. 88)

The Systemic Accident Model . . . focuses on the whole [system], not [just] the parts. It does not help you much to just focus on human errors, for example, or an equipment failure, without taking into account the sociotechnical system that helped shape the conditions for people's performance and the design, testing and fielding of that equipment. (p. 90)

2) Particular attention is given here to *Guidelines for Preventing Human Error in Process Safety* (CCPS, 1994). Although process safety appears in the book's title, the first two chapters provide an easily read primer on human error reduction. The content of those chapters was influenced largely by personnel with safety management experience at a plant or corporate level.

3) Safety practitioners should view the following excerpts as generic and broadly applicable. They advise on where human errors occur, who commits them and at what level, the effect of organizational culture and where attention is needed to reduce the occurrence of human errors. These excerpts apply to organizations of all types and sizes.

It is readily acknowledged that human errors at the operational level are a primary contributor to the failure of systems. It is often not recognized, however, that these errors frequently arise from failures at the management, design or technical expert levels of the company. (p. xiii)

A systems perspective is taken that views error as a natural consequence of a mismatch between human capabilities and demands, and an inappropriate organizational culture. From this perspective, the factors that directly influence error are ultimately controllable by management. (p. 3)

Almost all major accident investigations in recent years have shown that human error was a significant causal factor at the level of design, operations, maintenance or the management process. (p. 5)

One central principle presented in this book is the need to consider the organizational factors that create the preconditions for errors, as well as the immediate causes. (p. 5)

Factors such as the degree of participation that is encouraged in an organization and the quality of the communication between different levels of management and the workforce will have a major effect on the safety culture. (p. 5)

4) Since "failures at the management, design or technical expert levels of the company" affect the design of the workplace and work methods (i.e., the operating system) it is logical to suggest that safety professionals focus on system improvement to attain acceptable risk levels rather than principally on affecting worker behavior.

5) Reason's (1997) book, *Managing the Risks of Organizational Accidents*, is a must-read for safety professionals who would like an education in human error reduction. Reason writes about how the effects of decisions accumulate over time and become the causal factors for incidents that result in serious injuries or major damage when all the circumstances necessary for the occurrence of a major event fit together. This book stresses the need to focus on decision making above the worker level to prevent major incidents. Reason says:

> Latent conditions, such as poor design, gaps in supervision, undetected manufacturing defects or maintenance failures, unworkable procedures, clumsy automation, shortfalls in training, less than adequate tools and equipment, may be present for many years before they combine with local circumstances and active failures to penetrate the system's layers of defenses.
>
> They arise from strategic and other top-level decisions made by governments, regulators, manufacturers, designers and organizational managers. The impact of these decisions spreads throughout the organization, shaping a distinctive corporate culture and creating error-producing factors within the individual workplaces. (p. 10)

6) If the decisions made by management and others have a negative effect on an organization's culture and create error-producing factors in the workplace, focusing on reducing human errors at the worker level (unsafe acts) will not address the problems.

Prevention Through Design

Figure 2 depicts the prevention through design concept. Its intent is to display the advantages of moving prevention through design concepts upstream in the design process. Implementation of safety requirements are more easily achieved at that level and the cost of safety implementation is less than if retrofitting is necessary in the operation and maintenance mode. Overall, the premise is that hazards and risks are dealt with more effectively and economically in the design process.

1) Deming is right. Most of the problems in an operation are systemic, deriving from the workplace and work methods created by management. Responsibility for only the relatively small remainder lies with the workers.

2) Thus, companies can make great strides forward with respect to all hazard-related endeavors through the design and redesign processes.

3) For the practice of safety, the term *design and redesign processes* applies to facilities, hardware, equipment, tooling, selection of materials, operations layout and configuration, and work methods and procedures, personnel selection standards, training content, management of change procedures, maintenance requirements and personal protective equipment needs.

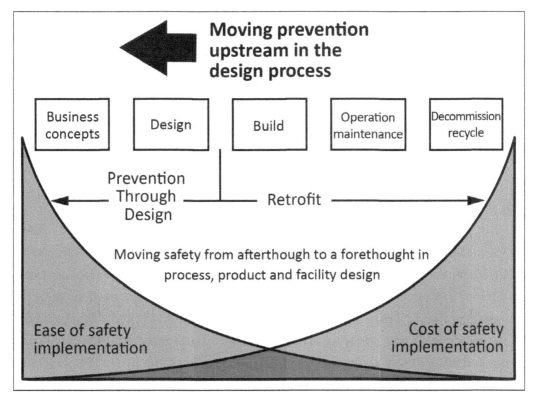

Figure 2: Prevention Through Design
Adapted from *Safety Through Design*, by W.C. Christensen and F.A. Manuele, 2000, Itasca, IL: National Safety Council.

4) The design and redesign processes aim to achieve acceptable risk levels.

5) Design and engineering applications that determine the workplace and work methods are the preferred measures of prevention since they are more effective in avoiding, eliminating and controlling risks.

6) Over time, the level of safety achieved will relate directly to the caliber of the initial design of the workplace and work methods, and their subsequent redesign in a continuous improvement endeavor.

7) A fundamental design goal, theoretically, is to have processes that are error proof. In *Quality Planning and Analysis*, Juran and Gryna (1983) speak appropriately of "error proofing the process":

> An important element of prevention is the concept of designing the process to be error free through "error proofing" (the Japanese call it *pokayoke* or *bakayoke*). A widely used form of error proofing is the design (or redesign) of the machines and tools (the "hardware") so as to make human error improbable or even impossible. (p. 347)

8) A company can usually meet the requirements to achieve an acceptable risk level in the design and redesign processes without great cost if the decision making takes place early enough upstream. When that does not occur and retrofitting to eliminate or control hazards is proposed, the cost may be prohibitive.

9) Safety practitioners should be aware of a reference on prevention through design found in ANSI/ASSE Z590.3-2011(R2016), Prevention Through Design: Guidelines for Addressing Occupational Hazards and Risks in Design and Redesign Processes.

On System Safety

1) ASSE's "Scope and Functions of the Professional Safety Position" says that the safety professional is to anticipate, identify and evaluate hazardous conditions and practices, and develop hazard control designs, methods, procedures and programs. Those are valid statements.

2) If safety professionals are to anticipate hazards, they must participate in the design processes. To be involved in the design process effectively, they must be skilled in hazard analysis and risk assessment techniques. Participating in the design processes and using hazard analysis and risk assessment techniques to achieve acceptable risk levels are the basics of system safety.

3) Applied system safety requires a conscientious, planned, disciplined and systematic use of special engineering and management tools on an anticipatory and forward-looking basis.

4) As stated in *The Loss Rate Concept in Safety Engineering,* Browning's (1980) premise is sound: As every loss event results from the interactions of elements in a system, it follows that all safety is system safety.

5) A significant premise of system safety is that hazards are most effectively and economically anticipated, avoided or controlled in the initial design process.

6) For workplace design, management and operations, and the task performance aspects of safety, application of hazard analysis and risk assessment methods are vital to achieving acceptable risk levels.

7) In *System Safety for the 21st Century,* Stephans (2004) makes this sensible statement:

> The safety of an operation is determined long before the people, procedures and plant and hardware come together at the work site to perform a given task. (p. 13)

This statement is valid and easily supported. Consider the thousands of decisions made by designers for a new facility that affect the level of safety achieved before people are hired for operations.

Setting Priorities and Utilizing Resources Effectively

1) These principles are postulated.

•All hazards do not present equal potential for harm or damage.

•All incidents that may result in injury, illness or damage do not have equal probability of occurrence, nor will their adverse outcomes be equal.

•Some risks are more significant than others.

•Resources are always limited. Staffing and money are never adequate to attend to all risks.

•The greatest good to employees, employers and society is achieved if available resources are effectively and economically applied to avoid, eliminate or control hazards and the risks that derive from them.

2) Since resources are always limited and since some risks are more significant than others, safety professionals must be able to distinguish the more significant from the less significant.

3) The professional practice of safety requires identifying the potentials for the greatest harm or damage for the decision makers and applying a ranking system to proposals made to avoid, eliminate or control hazards.

4) Safety practitioners must, therefore, be able to use hazard analysis and risk assessment methods of rating risks.

5) Causal factors for low-probability incidents resulting in severe harm or damage may be different from the causal factors for incidents that occur more frequently. Such low-probability incidents often involve unusual or nonroutine work, nonproduction activities, sources of high energy and certain construction situations.

6) Thus, safety professionals must undertake a separate and distinct activity to seek those hazards that present the most severe injury or damage potential so that they can be given priority consideration.

On Incident Causation

1) For most hazards-related incidents, even those that seem to present the least complexity, multiple causal factors may derive from less-than-adequate workplace and work methods design, management and operations, and personnel task performance.

2) In *MORT Safety Assurance Systems*, Johnson (1980) writes succinctly about the multifactorial aspect of incident causation:

> Accidents are usually multifactorial and develop through relatively lengthy sequences of changes and errors. Even in a relatively well-controlled work environment, the most serious events involve numerous error and change sequences, in series and parallel. (p. 74)

3) In the hazards-related incident process, deriving from those multiple causal factors:

• There are unwanted energy flows or exposures to harmful environments.

• A person or thing in the system, or both, are stressed beyond the limits of tolerance or recoverability.

• The incident process begins with an initiating event in a series of events.

• Multiple interacting events may occur, sequentially or in parallel, over time and influencing each other, to a conclusion that is likely to result in injury or damage.

4) Severity potential should determine whether hazards-related incidents are considered significant, even though serious harm or damage did not occur.

5) Heinrich has had more influence on the practice of safety than any other author. His premises have been adopted by many as certainty. They permeate the safety literature. Four editions of his book, *Industrial Accident Prevention,* were printed, the last being in 1959. Many of the Heinrich premises are questionable.

6) Heinrich's 88-10-2 ratios indicate that among the direct and proximate accident causes, 88% are unsafe acts, 10% are unsafe mechanical or physical conditions and 2% are unpreventable.

• The methodology used in arriving at those ratios cannot be supported.

• Current causation knowledge indicates the premise to be invalid.

• Heinrich's 88-10-2 premise conflicts with the work of others, including W. Edwards Deming, whose research finds that root causes derive from shortcomings in the management systems.

• Among the Heinrichean premises, application of the 88-10-2 ratios has had the greatest impact on the practice of safety, and has also done the most harm, since these ratios promote preventive efforts being focused on the worker rather on improving the operating system.

• Those who continue to promote the idea that 88% of all industrial incidents are caused primarily by the unsafe acts of persons do the world a disservice.

7) Heinrich's Foundation of a Major Injury, the 300-29-1 ratios (Heinrich's triangle), is the least tenable of his premises.

• It is impossible to conceive of incident data being gathered through the usual reporting methods in 1926 (when his postulation was made) in which 10 out of 11 reports would pertain to incidents that resulted in no injury.

• Conclusions pertaining to the 300-29-1 ratios were revised from one edition of his book to the next, without explanation, thus presenting questions about which version is valid.

- Heinrich's often-stated belief that the predominant causes of no-injury incidents are identical to the predominant causes of incidents resulting in major injuries is not supported by convincing statistical evidence and is questioned by several authors.
- Application of the premise results in misdirection since those who apply it may presume, inappropriately, that if they concentrate on reducing the types of events that occur frequently, they will also address the potential for severe injury.

8) Investigation of numerous incidents resulting in fatality or serious injury indicates that their causal factors are different and that they may not be linked to the causal factors for incidents that occur frequently and result in minor injury.

9) No documentation exists to support Heinrich's 4:1 ratio of indirect to direct injury costs. Further, arriving at a ratio that is applicable universally is implausible.

10) Heinrich places an inordinate emphasis on the unsafe acts of individuals as causal factors, while giving insignificant attention to systemic causal factors. This author believes that many safety practitioners would not agree with Heinrich's premise that "man failure is the heart of the problem and the methods of control must be directed toward man failure" (p. 4).

11) Heinrich gives prominence to causal factors that derive from ancestry and environment factors and to the faults of persons that allegedly derive from inherited or acquired faults (p. 15). That is inappropriate with respect to current societal mores.

12) Incident investigation, initially, should address the work system, applying a concept that:
- commences with inquiries to determine whether causal factors derive from workplace design decisions;
- examines whether the design of the work methods was overly stressful or error-provocative, or whether the immediate work situation encouraged riskier actions than the prescribed work methods.

Performance Measures

1) If the practice of safety professionals is based on sound science, engineering and management principles, it follows that safety professionals should be able to provide measures of performance that reflect the outcomes of the risk management initiatives they propose with some degree of accuracy. That has proven difficult to do.

2) Understanding the validity and shortcomings of the performance measures used is an indication of the maturity of the practice of safety as a profession.

3) Safety professionals must understand that the quality of the management decisions made to avoid, eliminate or control hazards and the risks that derive from them are affected directly by the validity of the information they provide. Their ability to provide accurate information to be used in decision making is a measure of their effectiveness.

4) Since an organization's safety achievements are a direct reflection of its culture, and since it takes a long time to change a culture, short-term performance measures should be examined cautiously as to validity.

5) Except for low-probability/high-consequence incidents, as the exposure base represented by the number of hours worked increases, the historical incident record has an increasing degree of confidence as a measure of the quality of safety in place, as well as a general, but not hazard-specific, predictor of future experience.

6) No statistical, historical performance measurement system can assess the quality of safety in place that encompasses low-probability/high-consequence incidents since such events seldom appear in the statistical history. For example, a risk assessment concludes that a defined catastrophic event, one that has not happened and is not represented in the statistical base, has an occurrence probability of once in 200 plant operating years.

7) Even for a large organization with significant annual hours worked, in addition to historical data, hazard-specific and qualitative performance measures (e.g., risk assessments, safety audits, perception surveys, incident recall technique) are also necessary, particularly to identify low-probability/severe-consequence risks.

8) Statistical process controls (e.g., cause-and-effect diagrams, control charts), as applied in quality management, can serve as performance measures for safety if the data set is large enough and if these controls are used prudently and with caution.

9) Incidents resulting in severe injury or damage seldom occur and would rarely be included on a statistical process control chart. Although such a chart may indicate that a system is in control, it could be deluding if it was presumed that the likelihood of low-probability incidents that could result in severe harm or damage was encompassed in the data.

10) It has been determined that having achieved stellar OSHA-type incident statistics does not ensure that controls are adequate with respect to the potential for having serious injuries or fatalities.

11) Since the language of management is finance, safety practitioners must be able to communicate incident experience in financial terms.

12) Much interest surrounds leading and lagging indicators to measure safety performance. Statistics traditionally gathered are lagging indicators. Leading indica-

tors include those such as training programs conducted, inspections performed and hazard communication sessions held. In the long term, management will still want to know whether application of the leading indicators has been successful, and the success will be measured largely by trends in trailing indicators, the incident experience and costs.

On Safety Audits

1) Safety audits must meet this definition to be effective: A safety audit is a structured approach to provide a detailed evaluation of safety effectiveness, a diagnosis of safety problems, a description of where and when to expect trouble, and guidelines concerning what should be done about the problems.

2) The paramount goal of a safety audit is to favorably influence the organization's culture. In *Safety Auditing: A Management Tool*, Kase and Wiese (1990) conclude properly that:

> Success of a safety auditing program can only be measured in the terms of the change it effects on the overall culture of the operation and enterprise that it audits. (p. 36)

3) Since evidence of an organization's culture and its management commitment to safety is first demonstrated through its upstream design and engineering decisions, safety audits that do not evaluate the design processes are incomplete and fall short of the definition of an audit.

4) Safety audits must also properly measure management commitment, primary evidence of which is a results-oriented accountability system. If such an accountability system does not exist, management commitment is questionable.

5) Safety audits must also determine whether adequate resources are provided to achieve and maintain acceptable risk levels.

Conclusion

It was said that this article would not contain a complete list of the principles for the practice of safety, knowing that others could add to it. However, the intent is to produce a document that encourages dialogue. To some extent, that has occurred and continues. ■

References

ANSI/AIHA/ASSE. (2012). Occupational health and safety management systems (Z10-2012). Retrieved from www.asse.org/departments/standards/safety_management

ANSI/ASSE. (2016). Prevention through design: Guidelines for addressing occupational hazards and risks in design and redesign processes [Z590.3-2011(R2016)]. Park Ridge, IL: ASSE.

ASSE. (1997). Scope and functions of the professional safety position. Des Plaines, IL: Author.

Brown, G.M., Hitchcock, D.E. & Willard, M.L. (1994). *Why TQM fails and what to do about it*. Burr Ridge, IL: Irwin Professional Publishing.

Browning, R.L. (1980). *The loss rate concept in safety engineering*. New York, NY: Marcel Dekker.

Center for Chemical Process Safety (CCPS). (1994). *Guidelines for preventing human error in process safety*. New York, NY: American Institute of Chemical Engineers.

Christensen, W.C. & Manuele, F.A. (2000). *Safety through design*. Itasca, IL: National Safety Council.

Dekker, S. (2006). *The field guide to understanding human error*. Burlington, VT: Ashgate Publishing Co.

Deming, W.E. (1986). *Out of the crisis*. Cambridge, MA: Center for Advanced Engineering Study, Massachusetts Institute of Technology.

EU-OSHA. (2017). Prevention and control strategies. OSHWiki. Retrieved from https://oshwiki.eu/wiki/Prevention_and_control_strategies

Fischhoff, B. (1989). *Risk: A guide to controversy*. Washington, DC: National Academies Press.

Grimaldi, J.V. & Simonds, R.H. (1989). *Safety management*. Homewood, IL: Irwin.

Heath, E.D. & Ferry, T. (1990). *Training in the workplace: Strategies for improved safety and performance*. Goshen, NY: Aloray.

Heinrich, H.W. (1959). *Industrial accident prevention* (4th ed.). New York, NY: McGraw Hill.

Johnson, W.G. (1980). *MORT safety assurance systems*. New York, NY: Marcel Dekker.

Juran, J.M. & Gryna, F.M. (1983). *Quality planning and analysis*. New York, NY: McGraw-Hill.

Kase, D.W. & Wiese, K.J. (1990). *Safety auditing: A management tool*. New York, NY: John Wiley & Sons.

Manuele, F.A. (2001). *Innovations in safety management: Addressing career knowledge needs*. New York, NY: John Wiley & Sons.

Manuele, F.A. (2013). *On the practice of safety* (4th ed.). Hoboken, NJ: John Wiley & Sons.

Manuele, F.A. (2014). *Advanced safety management: Focusing on Z10 and serious injury prevention* (2nd ed.). Hoboken, NJ: John Wiley & Sons.

Reason, J. (1997). *Managing the risks of organizational accidents*. Burlington, VT: Ashgate Publishing Co.

Stephans, R.A. (2004). *System safety for the 21st century*. New York, NY: John Wiley & Sons.

Walton, M. (1986). *The Deming management method*. New York, NY: Putnam Publishing Co.

2 SEVERE INJURY POTENTIAL
ADDRESSING AN OFTEN-OVERLOOKED SAFETY MANAGEMENT ELEMENT

ORIGINALLY PUBLISHED FEBRUARY 2003

Some safety practitioners presume that efforts focused on the types of accidents which occur frequently will also encompass the types of accidents that result in severe injury or damage. That premise is a reflection of H.W. Heinrich's belief, stated in the first edition of *Industrial Accident Prevention: A Scientific Approach*, that "the predominant causes of no-injury accidents are identical with the predominant causes of accidents resulting in major injuries" (90). Heinrich carried this idea forward in the later editions of his book.

But a differing observation has been made by some safety professionals—that incidents resulting in severe injury or damage are, mostly, unique and singular events; that their causal factors are different than those for accidents that result in minor injury; and that preventing their occurrence requires special safety management techniques.

In "Occupational Injuries: Factors Associated With Frequency and Severity," D. Kriebel offers these observations about the relation between injury frequency and severity.

> A Model of Injury Severity: Safety researchers have generally ignored severity, perhaps because for many years it was believed that the seriousness of the consequences of an accident was essentially randomly determined (Heinrich 1959). However, the analysis of the data from 89 industries (studied) . . . shows that the frequency and average severity of the injuries in an industry are poorly correlated one to the other, and are essentially independently determined (212).

In *MORT Safety Assurance Systems*, William Johnson also implies that severity potential needs greater emphasis.

> Some safety professionals are overly concerned with winning awards for reductions in minor injuries and underemphasize sources of serious accidents and disasters. Such a tendency can mislead management (19).

Can Serious Injury Potential Be Identified?

To a considerable degree, the answer to this question is yes. One can identify the types of work which produce many accidents that result in serious injury; then, the relevant hazards in that work can be addressed on an anticipatory basis. Although data in support of this premise are limited, it is persuasive. For example, in the second edition of *Safety Management*, Dan Petersen writes:

> If we study any mass data, we can readily see that the types of accidents that result in temporary total disabilities are different from the types of accidents resulting in permanent partial disabilities or in permanent total disabilities or fatalities. The causes are different. There are different sets of circumstances surrounding severity. Thus, if we want to control serious injuries, we should try to predict where they will happen. Today, we can often do just that.
>
> Studies in recent years suggest that severe injuries are fairly predictable in certain situations. Some of those situations involve:
> •unusual, nonroutine work;
> •nonproduction activities;
> •sources of high energy;
> •certain construction situations.
>
> These are just a beginning point. A long list could be made which would more extensively specify the areas where severity is predictable (11).

Data, not yet published and provided by Franklin Mirer, director of the Health and Safety Dept. at the International Union-UAW, is in concert with Petersen's observations. The data indicate that severe injury accidents occur disproportionately in unusual and nonroutine work, in nonproduction activities and where high sources of energy are present. During the 18 years prior to Jan. 1, 2002, the data show that skilled trades people—who represent about 20 percent of the UAW work population of about 700,000—make up 40 to 50 percent of the fatalities. These individuals include maintenance personnel, millwrights, electricians, steamfitters and tinsmiths; they rarely engage in repetitive and routine work, nor do they engage in production work. Total hours worked by the UAW population over this period is in the billions, which represents a sound statistical base; these data have significance.

The safety director of a chemical company notes that when the system is running, the risks are lower; when the system must be opened for maintenance or equipment fails or a chemical release occurs, severity potential is greatly increased. A safety director for a heavy electrical equipment manufacturer says that severe injuries in his company rarely occur during routine production operations.

However, severe injury or damage potential does exist in routine production work, and that potential can also be identified. The purpose here is not to suggest diminishing efforts to prevent accidents that produce less-than-severe injuries. Rather, the intent is to encourage adoption of additional efforts to prevent accidents that result in severe injury.

Cascading Events: Accidents That Result in Serious Injury

What have others said about the nature of incidents that result in serious injury? In *The Psychology of Everyday Things*, Donald Norman asserts that "it is spectacularly easy to find examples of false assessment in industrial accidents." Norman teaches undergraduate and graduate classes entitled "Cognitive Engineering" at the University of California, San Diego. He writes:

> Explaining away errors is a common problem in commercial accidents. Most major accidents follow a series of breakdowns and errors, problem after problem, each making the next more likely. Seldom does a major accident occur without numerous failures: equipment malfunctions, unusual events, a series of apparently unrelated breakdowns and errors that culminate in major disaster; yet no single step has appeared to be serious. In many of these cases, the people involved noted the problem but explained it away, finding a logical explanation for the otherwise deviant observation (128).

Note the terminology "numerous failures" and "a series of apparently unrelated breakdowns and errors." An aspect of many incidents that result in severe injury is the cascading effect of multiple causal factors acting in sequence—sometimes in parallel sequences—toward an undesirable end. Kingsley Hendrick and Ludwig Benner Jr. offer similar comments about the cascading effect of events in accident occurrences in *Investigating Accidents With STEP*.

> Because accidents are composed of sets of individual events, all of which are interrelated, each event affects one or more actors and what they do next, changing their state. The first event in the accident process is a perturbation or an undesired or unplanned change in someone or something within the planned process. That first disruptive event initiates a sort of cascading effect, culminating in some harm or loss (31).

The term "actor" is used to identify the people or things who or which directly influenced the flow of events that construct the accident sequence (69). STEP is an acronym for "sequentially timed events plotting." It is an events-analysis-based approach in which events are plotted sequentially (and in parallel, if appropriate) to show the cascading effect as each event impacts on others. It is built on the management system embodied in the management oversight and risk tree (MORT) and system safety technology.

Having been involved in many incident investigations and having reviewed thousands of accident reports, the author concludes that many accidents which result in severe injury are unique and singularly occurring events in which a series of breakdowns occur in a cascading effect. That phenomenon calls for creating and implementing methods that identify serious injury potentials and mitigate against their occurrence. Preventing such accidents requires a strategy that specifically addresses serious injury potential in every aspect of safety management—from initial design through dismantling and disposition.

Risk Avoidance and Control in the Design Process

Avoiding serious injury potential is most effectively accomplished in the original design process, although redesign activities (applying the same safety through design principles) present similar opportunities when retrofitting occurs. The auto industry provides an example of methods to be used to avoid injury potential— particularly serious injury potential—both in the initial design and subsequent redesign processes. Although the following excerpt is taken from the 1999 General Motors/UAW labor agreement, similar language appears in such contracts with other auto companies.

> As early as possible and preferably in the zero phase of the planning in the design process . . . the parties agree to perform task-based risk assessments on new equipment and manufacturing systems, and on existing equipment and manufacturing systems where locally agreed to and approved by the Plant Safety Review Board. A task-based risk assessment will be performed after the detailed designs are completed. . . . A review of anticipated equipment and/or processes with the shop committee and the Local Joint Health and Safety Committee will be held.
>
> The Local Joint Health and Safety Committee may be required to travel to vendors, plants or other locations to participate in a design review of such equipment or processes as outlined in the Design for Health and Safety Specification.
>
> Machinery, equipment or processes will not be released for production without the written approval of the plant safety administrator.

In summary, this agreement presents a theoretical ideal. If it became a model, universally applied, serious injury potential would be significantly reduced.

A Relevant European Guideline:
Impacting Design and Operations Considerations

With respect to guidelines and standards that include provisions applicable to the prevention of accidents which result in severe injuries or fatalities, this author believes that the Europeans are the world's leaders. Little safety literature applies specifically to severe injury potential. One exception is "Guidelines on a Major Accident Prevention Policy and Safety Management System as Required by Council Directive 96/82/EC (Seveso II)" (which can be found at http://mahbsrv.jrc.it/NewProducts-SafetyManagementSystems.html).

This document was issued by the European Commission-Joint Research Centre, Institute for Systems Information and Safety, Major Accident Hazards Bureau. It reflects the intent of Council Directive 96/82/EC (SEVESO II) which "is aimed at the prevention of major accidents involving dangerous substances and the limitation of their consequences." The document features four major sections: introduction to safety management systems; development of major accident prevention policy; elements of safety management systems; and bibliography.

In particular, the section on elements of safety management systems speaks to the topic at hand. It contains these subsections: organization and personnel; hazard identification and evaluation; operational control; management of change; planning for emergencies; monitoring performance; and audit and review. While all of these elements are significant in avoiding severe-injury-producing accidents, the subsection on hazard identification and evaluation is particularly relevant. It requires hazard and risk identification and avoidance or mitigation, both on an anticipatory basis in the design process and during all phases of operations. It reads:

> The following issues shall be addressed by the safety management system (SMS): identification and evaluation of major hazards—adoption and implementation of procedures for systematically identifying major hazards arising out of normal and abnormal operation and the assessment of their likelihood and severity.

The following excerpts, taken from the text that follows the citation above, specify the actions to be taken with respect to major hazards.

> Hazard identification and evaluation procedures should be applied to all relevant stages from project conception through to decommissioning, including:
> •potential hazards arising from or identified in the course of planning, design, engineering, construction, commissioning and development activities;
> •the normal range of process operating conditions, hazards of routine operations and nonroutine situations, in particular start-up, maintenance and shut down.

Thus, these requirements encompass identifying hazards and risks both on an anticipatory basis in the design process and in the entire spectrum of operations. That is vital in minimizing severe injury potential. If no hazards are present—no potential for harm—no accidents can occur.

A companion piece to the guidelines is the book *Prevention of Major Industrial Accidents*, which is an International Labor Office "code of practice"; its content parallels the guidelines, but is more extensive.

OSHA and EPA

Although OSHA's Process Safety Management (PSM) standard and EPA's Risk Management Program (RMP) regulation could be considered to address severe injury potential, their terminology is not specifically so directed. OSHA 1910.119, which pertains to PSM of highly hazardous chemicals, contains provisions such as process hazard analysis, operating procedures, prestartup safety review, management of change, and emergency planning and response. Those provisions can be interpreted as serious injury prevention measures.

While OSHA's regulatory authority pertains to on-site consequences, EPA's concerns center on offsite consequences. EPA 40 CFR Part 68 mandates RMPs, which are designed to help prevent accidental chemical releases. It contains provisions for hazard reviews and control, and overlaps considerably with the PSM requirements. However, neither rule includes language similar to that found in the European guidelines with respect to potential hazards arising from or identified in the course of planning, design, engineering, construction, commissioning and development activities.

Other Prevention Techniques

In addition to the cited methods to avoid serious injury, other preventive techniques are available. Although the suggestion here is that the following techniques be adapted as specific measures to identify hazards that present serious injury potential, the methods can encompass the prevention of all types of incidents. The techniques are the critical incident technique and prejob planning for nonroutine work.

The Critical Incident Technique

The critical incident technique is used to identify and take action on hazards that pose serious injury potential. Skilled observers interview a sampling of personnel, eliciting their recall of "critical" incidents which have exposed them to operational or physical hazards that caused them concern, whether or not injury occurred. For this process to succeed, those involved must recognize that workers being interviewed are a valuable resource in identifying hazards and risks because of their extensive knowledge of how the work is performed.

Critical incidents identified are analyzed and classified with respect to the significance of the risks presented by hazards identified, with priorities set for remedial action (Grimaldi and Simonds 248; NSC 101; Tarrants; Johnson 386). Johnson's comments on this technique are of particular interest. With respect to incident recall he writes:

> Incident recall is an information gathering technique whereby employees (participants) describe situations they have personally witnessed involving good and bad practices and safe and unsafe conditions.
>
> Such studies, whether by interview or questionnaire, have a proven capacity to generate a greater quantity of relevant, useful reports than other monitoring techniques, so much so as to suggest that their presence is an indispensable criterion of an excellent safety program (386).

Prejob Planning and Safety Reviews for Nonroutine Work

Based on the author's experience and review of incident reports on severe injuries resulting from accidents that occurred during nonroutine work, it can be said (at least anecdotally) that the work likely would not have been performed the way it was had a prejob planning and safety review been conducted. And, while further inquiry suggests that prejob safety reviews are not the norm in many companies, interest appears to be growing. Based on a study in Michigan, a report titled, "Risk Assessment for Maintenance Activities: Preventing Injuries Before They Happen" was issued in October 2001 and made available via the Internet in March 2002 (Main). In the two weeks following its posting, 375 downloads were recorded. The maintenance personnel included in the survey usually perform nonroutine work—and as noted, they experience a disproportionate share of the serious injuries and fatalities. Therefore, it appears that conducting prejob safety reviews for nonroutine work would greatly reduce the risk of severe injury and fatality.

It should be noted that establishing the concept is what is most important. Care must be taken to not create extensive procedures and reports when that is not necessary for a particular and simplistic nonroutine job. For many such jobs, it is sufficient if the work planning and safety review conducted are simply brief discussions that allow those involved to arrive at a go or no-go conclusion. Ideally, it will become standard practice for workers to think through the job to be performed and to plan for the methods to be used, discuss the hazards and risks, and determine whether the risks are acceptable. Of course, if they conclude that the risks posed are not acceptable, a more thorough job review and risk assessment will be necessary.

Achieving a culture change that incorporates prejob planning and safety analysis as an accepted and expected practice requires support from all levels of management and from workers. Such a change cannot be attained without training that helps workers understand the key concepts. As the following example demonstrates, this change can be achieved.

The Concept in Practice

At a large location, the severe injury experience was considered excessive for nonroutine work and needed to be addressed. As staff safety professionals prepared a course of action and talked it up with all personnel—from top management to hourly workers— they encountered the usual negatives—"Skilled tradespeople won't buy into the program"; "Skilled trades supervisors resist any change"—which were viewed as expressions of normal resistance to change.

In effect, the program consisted of indoctrinating management and the workforce in the benefits of conducting a prejob review that encompassed how to complete the job effectively in a timely manner, as well as job hazard analysis and risk assessment. Eventually, management and skilled trades personnel agreed that information sessions would be held (which the safety professionals later called vital to their success).

At the beginning of those sessions, attendees received a discussion outline that presented the fundamentals of the proposed prejob review system. After discussing the outline, attendees were divided into groups to plan actual maintenance jobs described in various scenarios. Following is a comparable outline that is a composite of prejob planning and safety analysis methods.

Prejob Planning and Safety Analysis Outline

1) Review the scenario that defines the work to be done. Considering both safety and productivity:
   a) Break down the job into manageable tasks.
   b) Determine:
   • how each task will be performed;
   • in what order the tasks will be performed;
   • what equipment or materials are needed;
   • whether any particular skills are necessary.
2) Will the work require: a hot work permit, a confined space entry permit, lockout/tagout (of what equipment or machinery)?
3) Will it be necessary to barricade for clear work zones?
4) Will aerial lifts be required?
5) What PPE will be required?
6) Will fall protection be required?
7) What are the hazards in each task? Consider:

Fire	Explosion	Pressure
Work at heights	Work at depth	Vibration
Pinch points	Fall hazards	Electricity
Chemicals	Dusts	Noise
Weather	Sharp objects	Steam
Elevated loads	Stored energy	Dropping tools
Moving equipment	Forklifts	Hot objects
Conveyors	Access	Weight

These are examples only. A hazards list should be developed to suit the hazards and risks inherent in the operation.

8) Of the hazards identified, do any present severe risk of injury?
9) Develop hazard control measures, applying the safety decision hierarchy.
- Eliminate hazards and risks through system design and redesign.
- Reduce risks by substituting less-hazardous methods/materials.
- Incorporate safety devices (fixed guards, interlocks).
- Provide warning systems.
- Apply administrative controls (work methods, training, etc.).
- Provide personal protective equipment.

10) Is any special contingency planning necessary? People? Procedures?
11) What are workers to do if the work doesn't go as planned?
12) Considering the foregoing, are the risks acceptable? If not, what action should be taken?

At this location, skilled trades supervisors became proponents of prejob analysis and planning once they recognized that it made their jobs easier, improved productivity and reduced risks. As one safety professional involved says, "Our skilled trades supervisors who have been involved in the process have become real believers in it." A culture change had been achieved.

Some companies require contractors that perform work on their premises to submit written prejob safety plans as part of the bid qualification process. In that respect, the construction industry is ahead of other business and industry categories. Figure 1 presents a prejob plan and safety analysis prepared before work commenced on a construction project.

Conclusion

To a large extent, hazards and risks that present severe injury or fatality potential are identifiable. Preferably, these hazards/risks would be addressed in the design processes for facilities, equipment, operating systems, tooling, processes or products, and in the design of the work methods. Data on fatalities and serious injuries establishes that hazards and risks with severity potential are not always considered and reduced to a practical minimum in the design process, however. Therefore, it is not unusual to find such potential in facilities and operations.

Unfortunately, some safety practitioners continue to act on the premise that if efforts are concentrated on frequently occurring accidents, the potential for severe injury will also be addressed. That results in severe injury potential being overlooked, since the types of accidents that produce severe injuries or fatalities are rarely represented in the data pertaining to accidents which occur frequently. A sound case can be made that many accidents which result in severe injury or fatality are unique and

Work to be performed:	Install siding on the outside perimeter of the welding shop using a 40-ft. aerial work platform
Analysis prepared by:	Irene Dunne, General Supervisor Date: 6/5/2002
Competent Person:	T. Paine Qualified Person: H. Lee

Hazard	Safe Work Practices & Hazard Control	Contingency Plan
Falling from aerial work platform.	• Employees will utilize full-body harness with shock-absorbing lanyard attached to manufacturer's identified anchorage point in the aerial work platform. • Employees will be trained to use the aerial work platforms used on this project. Hard hat stickers will be issued and displayed on the hard hats of those trained. • At the beginning of each shift, the aerial work platform operator will perform a visual inspection and functional test according to the manufacturer's recommendation. A copy of the inspection and test will remain with the equipment for the entire shift.	Building phones are not available. When using a cell phone, the **emergency number is (123) 456-7890.**
Aerial work platform tipover (adjacent to an area of excavation work). Muddy ground conditions due to poor drainage may cause problems.	• Due to heavy rain and poor drainage it will be necessary to monitor and test the soil conditions and terrain surrounding the building in each new area and before work begins to ensure a safe foundation for the aerial work platform and mobile crane operations. Stable conditions will be provided, where necessary, through use of mudsill blocks, cribbing, or other acceptable means for effective wheel contact, outrigger placement and equipment support. • **T. Paine** will also be on the ground, rigging and attaching the siding panels. • **T. Paine** will inspect ground conditions on a daily basis and after any rain, paying close attention to areas around nearby excavations. • **T. Paine** will inspect the path of travel and ground conditions in each new area where the aerial work platform will be set up to install siding.	If workers in the aerial work platform are unable to operate the controls, the **competent** ground person, **T. Paine,** will begin a rescue by operating the ground controls. T. Paine is also trained in first aid and CPR.
Falling sheets of siding, materials or tools.	• Siding panels will be hoisted into fastening position by using a mobile crane and proper rigging technique. • Siding panels will be securely fastened before releasing the hoisting apparatus.	
Electrical shock.	• All electrical power tools and extension cords will be GFCI protected.	
People entering into the hazard work area.	• Portable 42-in. barricades will be used to deter entry of unauthorized personnel into the hazard area.	
Inclement weather conditions.	• Consideration will be given to weather-related conditions. If the wind, rain or storm conditions, or other situations regarding hazardous weather occur, this operation will be temporarily suspended.	
Persons falling from roof.	• Employees working on the roof will utilize full-body harnesses with shock-absorbing lanyards attached to the existing anchorage point(s). The capacities were verified by our P.E.	The qualified person for fall hazard control on this job is **H. Lee.**

Figure 1: Example Prejob Plan and Safety Analysis

singular events, yet safety management systems rarely specifically address severe injury potential. Thus, to properly address that potential, safety practitioners must undertake separate and distinct activities to identify hazards which present severe injury or damage potential, so that they can be given the priority consideration needed. ■

References

EPA. "General Guidance for Risk Management Programs." 40 CFR Part 68, Chemical Accidental Release Prevention. Washington, DC: EPA, July 1998.

Grimaldi, J.V. and R.H. Simonds. *Safety Management*. 5th ed. Homewood, IL: Irwin Press, 1989.

"Guidelines on a Major Accident Prevention Policy and Safety Management System." As required by Council Directive 96/82/EC. Luxembourg: Office for Official Publications of the European Communities, 1998.

Hendrick, K. and L. Benner Jr. *Investigating Accidents With STEP*. New York: Marcel Dekker Inc., 1987.

Heinrich, H.W. *Industrial Accident Prevention: A Scientific Approach*. 1st ed. New York: McGraw-Hill, 1931.

International Labor Office (ILO). *Prevention of Major Accidents*. Geneva, Switzerland: International Occupational Safety and Health Information Center, ILO, 1993.

Johnson, W.G. *MORT Safety Assurance Systems*. New York: Marcel Dekker Inc., 1980.

Kriebel, D. "Occupational Injuries: Factors Associated with Frequency and Severity." *International Archives of Occupational and Environmental Health*. 50(1982): 209-218.

Main, B.W. "Risk Assessment for Maintenance Activities: Preventing Injuries Before They Happen." Ann Arbor, MI: Design Safety Engineering, 2001.

National Safety Council (NSC). *Accident Prevention Manual: Administration and Programs*. 12th ed. Itasca, IL: NSC, 2001.

Norman, D.A. *The Psychology of Everyday Things*. New York: Basic Books, 1998.

OSHA. "Process Safety Management of Highly Hazardous Chemicals." 29 CFR 1910.119. Washington, DC: U.S. Dept. of Labor, OSHA, 1992.

Petersen, D. *Safety Management*. 2nd ed. Des Plaines, IL: ASSE, 1998.

Tarrants, W.E., ed. *The Measurement of Safety Performance*. New York: Garland Publishing, 1980.

3 Is a Major Accident About to Occur in Your Operations?
Lessons to Learn from the Space Shuttle *Columbia* Explosion

ORIGINALLY PUBLISHED MAY 2004

The report issued in August 2003 by the Columbia Accident Investigation Board (CAIB) is revealing about how a lot of little things can add up to a big thing. This report provides SH&E practitioners a basis for reflection on the potential for the occurrence of major accidents in their operations. This article is presented in three parts. Part 1 reviews the origins of causal factors for accidents that result in serious consequences. Part 2 presents excerpts from Volume 1 of CAIB's report. Part 3 presents a discussion guide which can be used to determine whether latent hazardous conditions and practices that could be the causal factors for a major accident have accumulated in a given setting. This guide provides the basis for a cultural, organizational and technical self-evaluation. It is for use by SH&E practitioners and, more particularly, for the executives influenced by those practitioners to undertake a review of major accident potential.

Part 1: Causal Factors
Major accidents—meaning low-probability incidents with severe consequences—typically result from an accumulation of what Reason refers to as latent conditions. Such latent technical conditions and operating practices are built into a system and shape an organization's culture. In *Managing the Risks of Organizational Accidents*, Rea-

son discusses the long-term impact of a continuum of less-than-adequate safety decision making—which is a central theme in CAIB's report:

> Latent conditions, such as poor design, gaps in supervision, undetected manufacturing defects or maintenance failures, unworkable procedures, clumsy automation, shortfalls in training, less than adequate tools and equipment, may be present for many years before they combine with local circumstances and active failures to penetrate the system's layers of defenses. They arise from strategic and other top-level decisions made by governments, regulators, manufacturers, designers and organizational managers. The impact of these decisions spreads throughout the organization, shaping a distinctive corporate culture and creating error-producing factors within the individual workplaces (10).

In this paragraph, Reason cites many of the cultural and organizational shortcomings that resulted in less-than-adequate safety decision-making at NASA, which CAIB considered significant. In *The Psychology of Everyday Things,* Norman writes similarly about how major accidents occur:

> Explaining away errors is a common problem in commercial accidents. Most major accidents follow a series of breakdowns and errors, problem after problem, each making the next more likely. Seldom does a major accident occur without numerous failures: equipment malfunctions, unusual events, a series of apparently unrelated breakdowns and errors that culminate in major disaster; yet no single step has appeared to be serious. In many cases, the people noted the problem but explained it away, finding a logical explanation for the otherwise deviant observation (128).

What Norman says about "numerous failures" being typical when major accidents occur matches this author's experience, having reviewed many accident reports pertaining to severe injuries and fatalities. While reading the excerpts from CAIB's report later this article, two key quotes should be kept in mind:

> The impact of [top-level] decisions spreads throughout the organization, shaping a distinctive corporate culture and creating error-producing factors within individual workplaces (Reason 10).

> In many cases, the people noted the problem but explained it away, finding a logical explanation for the otherwise deviant observation (Norman 128).

SH&E professionals should review CAIB's entire investigation report, which is available at www.nasa.gov/columbia/home/CAIB_Vol1.html. While it is upsetting, readers are reminded that similar latent technical conditions and operating practices could exist in their operations.

The reality is that the following scenario is often repeated. A location does well for several years as measured by its safety statistics. Then, a major accident occurs and everyone is shocked that such an incident could happen in their operations. After all, wasn't the safety record commendable?

Unfortunately, what follows a major accident is well-described in CAIB's report.

> Many accident investigations do not go far enough. They identify the technical cause of the accident, and then connect it to a variant of "operator error"—the line worker who forgot to insert a bolt, the engineer who miscalculated the stress or the manager who made the wrong decision. But this is seldom the entire issue. When the determinations of the causal chain are limited to the technical flaw and individual failure, typically the actions taken to prevent a similar event in the future are also limited: Fix the technical problem and replace or retrain the individual responsible. Putting these corrections in place leads to another mistake—the belief that the problem is solved. The board did not want to make these errors. (CAIB 97).

SH&E practitioners who still profess that most work-related accidents are principally caused by unsafe acts of workers should seriously consider the report excerpts that follow. Perhaps their incident investigation procedures do not go far enough and should be extended to identify the real root-cause factors, as CAIB did.

Furthermore, it is believed that the highlights of this report form a base from which operations managers and SH&E practitioners can assess whether there have been shortcomings in decision making in the past with respect to safety in the operations they influence. Such an assessment could determine whether these shortcomings have resulted in an accumulation of latent conditions and operating practices that may have serious injury potential. It will also result in an assessment of the organization's safety culture.

The verbatim excerpts from the 248-page Volume 1 describe cultural deficiencies that may exist in any operation. They also reinforce several premises:

- Causal factors for accidents that result in severe injuries are multiple and complex, and relate to several levels of responsibility.
- Accident investigations often blame a failure only on the last step in a complex process, when a more comprehensive understanding of that process could reveal that earlier steps might be equally or even more culpable.
- Accidents that result in severe injuries may not be random events; rather, their causal factors may derive from an accumulation over time of deficiencies in an organization's safety culture.
- An organization's culture with respect to safety decision making determines the incident experience obtained.

Part 2: CAIB Report Excerpts
The Board Statement

Our aim has been to improve shuttle safety by multiple means, not just by correcting the specific faults that cost the nation this orbiter and this crew. With that intent, the board conducted not only an investigation of what happened to *Columbia*, but also to determine the conditions that allowed the accident to occur—a safety evaluation of the entire space shuttle program.

It is our view that complex systems almost always fail in complex ways, and we believe it would be wrong to reduce the complexities and weaknesses associated with these systems to some simple explanation.

In this board's opinion, unless the technical, organizational and cultural recommendations made in this report are implemented, little will have been accomplished to lessen the chance that another accident will follow (CAIB 6).

The Executive Summary

The board recognized early on that the accident was probably not an anomalous, random event, but rather likely rooted to some degree in NASA's history and the human space flight program's culture. Accordingly, the board broadened its mandate at the outset to include an investigation of a wide range of historical and organizational issues, including political and budgetary considerations, compromises and changing priorities over the life of the space shuttle program. The board's conviction regarding the importance of these factors strengthened as the investigation progressed, with the result that this report, in its findings, conclusions and recommendations, places as much weight on these causal factors as on the more easily understood and corrected physical cause of the accident (CAIB 9).

[Note: The executive summary remarks extensively on the physical and organizational causal factors for the accident. In-depth comments about the causal factors are found later in the report.]

The Report Synopsis

We consider it unlikely that the accident was a random event; rather it was likely related in some degree to NASA's budgets, history and program culture, as well as to the politics, compromises [and] changing priorities of the democratic process. We are convinced that the management practices overseeing the space shuttle program were as much a cause of the accident as the foam that struck the left wing (CAIB 11).

CAIB Report Part One: The Accident
Chapter 1: The Evolution of the Space Shuttle Program

It is the view of the Columbia Accident Investigation Board that the *Columbia* accident is not a random event, but rather a product of the space shuttle program's history and current management processes. Fully understanding how it happened requires an exploration of that history and management. This chapter charts how the shuttle emerged from a series of political compromises that produced unreasonable expectations—even myths—about its performance, how the *Challenger* accident shattered those myths several years after NASA began acting upon them as fact, and how, in retrospect, the shuttle's technically ambitious design resulted in an inherently vulnerable vehicle, the safe operation of which exceeded NASA's organizational capabilities as they existed at the time of the *Columbia* accident.

To understand the cause of the *Columbia* accident is to understand how a program promising reliability and cost efficiency resulted instead in a developmental vehicle that never achieved the fully operational status NASA and the nation accorded it (CAIB 21).

The *Challenger* Accident

When the Rogers Commission discovered that, on the eve of the launch, NASA and a contractor had vigorously debated the wisdom of operating the shuttle in the cold temperatures predicted for the next day, and that more senior NASA managers were unaware of this debate, the commission shifted the focus of its investigation to "NASA management practices, center-headquarters relationships and the chain of command for launch commit decisions." As the investigation continued, it revealed a NASA culture that had gradually begun to accept escalating risk, and a NASA safety program that was largely silent and ineffective (CAIB 25).

Chapter 3: Accident Analysis
The Physical Cause

The physical cause of the loss of *Columbia* and its crew was a breach in the thermal protection system on the leading edge of the left wing. The breach was initiated by a piece of insulating foam that separated from the left bipod ramp of the external tank and struck the wing in the vicinity of the lower half of reinforced carbon-carbon panel 8 at 81.9 seconds after launch. During re-entry, this breach . . . allowed superheated air to penetrate the leading-edge insulation and progressively melt the aluminum structure of the left wing, resulting in a weakening of the structure until increasing aerodynamic forces caused loss of control, failure of the wing and breakup of the orbiter (CAIB 49).

STS-107 Left Bipod Foam Ramp Loss

Foam loss has occurred on more than 80 percent of the 79 missions for which imagery is available, and foam was lost from the left bipod ramp on nearly 10 percent of missions where the left bipod ramp was visible following external tank separation (CAIB 53).

The precise reasons why the left bipod foam ramp was lost from the external tank during STS-107 [the *Columbia* mission] may never be known. The specific initiating event may likewise remain a mystery. However, it is evident that a combination of variable and pre-existing factors, such as insufficient testing and analysis in the early design stages, resulted in a highly variable and complex foam material, defects induced by an imperfect and variable application, and the results of that imperfect process, as well as severe load, thermal, pressure, vibration, acoustic, and structural launch and ascent conditions (CAIB 53, 55).

The Orbiter "Ran Into" the Foam

"How could a lightweight piece of foam travel so fast and hit the wing at 545 miles an hour?" Just prior to separating from the external tank, the foam was traveling with the shuttle stack at about 1,568 mph (2,300 feet per second). Visual evidence shows that the foam debris impacted the wing approximately 0.161 seconds after separating from the external tank. In that time, the velocity of the foam debris slowed from 1,568 mph to about 1,022 mph (1,500 feet per second). Therefore, the orbiter hit the foam with a relative velocity of about 545 mph (800 feet per second). In essence, the foam debris slowed down and the orbiter did not, so the orbiter ran into the foam. The foam slowed down rapidly because such low-density objects have low ballistic coefficients, which means that their speed rapidly decreases when they lose their means of propulsion (CAIB 60).

Orbiter Sensors

Nearly all of *Columbia*'s sensors were specified to have only a 10-year shelf life, and in some cases an even shorter service life. At 22 years old, the majority of the orbiter experiment instrumentation had been in service twice as long as its specified service life and, in fact, many sensors were already failing. Engineers planned to stop collecting and analyzing data once most of the sensors had failed, so failed sensors and wiring were not repaired. For instance, of the 181 sensors in *Columbia*'s wings, 55 had already failed or were producing questionable readings before STS-107 was launched (CAIB 65).

Findings

[During re-entry] abnormal heating events preceded abnormal aerodynamic events by several minutes. By the time data indicating problems was telemetered to Mission Control Center, the orbiter had already suffered damage from which it could not recover (CAIB 73).

CAIB Report Part Two: Why the Accident Occurred

In our view, the NASA organizational culture had as much to do with this accident as the foam. Organizational culture refers to the basic values, norms, beliefs and practices that characterize the functioning of an institution. At the most basic level, organizational culture defines the assumptions that employees make as they carry out their work. It is a powerful force that can persist through reorganizations and the change of key personnel. It can be a positive or a negative force.

At NASA's urging, the nation committed to building an amazing, if compromised, vehicle called the space shuttle. When the agency did this, it accepted the bargain to operate and maintain the vehicle in the safest possible way. The board is not convinced that NASA has completely lived up to the bargain, or that Congress and the administration have provided the funding and support necessary for NASA to do so. This situation needs to be addressed—if the nation intends to keep conducting human space flight, it needs to live up to its part of the bargain (CAIB 97).

Chapter 5: From *Challenger* to *Columbia*

The board is convinced that the factors that led to the *Columbia* accident go well beyond the physical mechanisms [previously] discussed. The causal roots of the accident can also be traced, in part, to the turbulent post-Cold War policy environment in which NASA functioned during most of the years between the destruction of *Challenger* and the loss of *Columbia*.

The agency could not obtain budget increases through the 1990s. Rather than adjust its ambitions to this new state of affairs, NASA continued to push an ambitious agenda of space science and exploration, including a costly space station program.

The space shuttle program has been transformed since the late 1980s implementation of post-*Challenger* management changes in ways that raise questions . . . about NASA's ability to safely operate the space shuttle. While it would be inaccurate to say that NASA managed the space shuttle program at the time of the *Columbia* accident in the same manner it did prior to *Challenger*, there are unfortunate similarities between the agency's performance and safety practices in both periods (CAIB 99).

Space Shuttle Program Budget Patterns

In Fiscal Year 1993, the outgoing Bush administration requested $4.128 billion for the space shuttle program; five years later, the Clinton administration request was for $2.977 billion, a 27-percent reduction. By Fiscal Year 2003, the budget request had increased to $3.208 billion, still a 22-percent reduction from a decade earlier. With inflation taken into account, over the past decade, there has been a reduction of approximately 40 percent in the purchasing power of the program's budget, compared to a reduction of 13 percent in the NASA budget overall (CAIB 104).

Conclusion

[T]his is hardly an environment in which those responsible for safe operation of the shuttle can function without being influenced by external pressures. It is to the credit of space shuttle managers and the shuttle workforce that the vehicle was able to achieve its program objectives for as long as it did. An examination of the shuttle program's history from *Challenger* to *Columbia* raises the question: Did the space shuttle program budgets constrained by the White House and Congress threaten safe shuttle operations? There is no straightforward answer. At the time of the launch of STS-107, NASA retained too many negative (and also many positive) aspects of its traditional culture: "flawed decision making, self deception, introversion and a diminished curiosity about the world outside the perfect place." These characteristics were reflected in NASA's less-than-stellar performance before and during the STS-107 mission (CAIB 118).

Chapter 6: Decision Making at NASA
A History of Foam Anomalies

The shedding of external tank foam—the physical cause of the *Columbia* accident—had a long history. Damage caused by debris has occurred on every space shuttle flight, and most missions have had insulating foam shed during ascent. This raises an obvious question: Why did NASA continue flying the shuttle with a known problem that violated design requirements? It would seem that the longer the shuttle program allowed debris to continue striking the orbiters, the more opportunity existed to detect the serious threat it posed. But this is not what happened (CAIB 121).

Original Design Requirements

Early in the space shuttle program, foam loss was considered a dangerous problem. Design engineers were extremely concerned about potential damage to the orbiter and its fragile thermal protection system, parts of which are so vulnerable to impacts that lightly pressing a thumbnail into them leaves a mark (CAIB 121).

Findings

Foam-shedding, which had initially raised serious safety concerns, evolved into "in-family" or "no-safety-of-flight" events or were deemed an "accepted risk" (CAIB 130).

NASA failed to adequately perform trend analysis on foam losses. This greatly hampered the agency's ability to make informed decisions about foam losses (CAIB 131).

Discovery and Initial Analysis of Debris Strike

In the course of examining film and video images of *Columbia*'s ascent, the Intercenter Photo Working Group identified, on the day after launch, a large debris strike to the leading edge of *Columbia*'s left wing. Alarmed at seeing so severe a hit so late in ascent, and at not having a clear view of damage the strike might have caused, Intercenter Photo Working Group members alerted senior program managers by phone and sent a digitized clip of the strike to hundreds of NASA personnel via e-mail. These actions initiated a contingency plan that brought together an interdisciplinary group of experts from NASA, Boeing and the United Space Alliance to analyze the strike. So concerned were Intercenter Photo Working Group personnel that on the day they discovered the debris strike, they tapped their chair . . . to see through a request to image the left wing with Dept. of Defense assets in anticipation of analysts needing these images to better determine potential damage. By the board's count, this would be the first of three requests to secure imagery of *Columbia* on-orbit during the 16-day mission (CAIB 166).

[Note: Thirty-two pages in Volume 1 are devoted to decision making pertaining to analysis of the initial foam strike. Under the caption "Missed Opportunities," the report discusses eight situations whereby management personnel might have decided to arrange for the requested imagery. Comparable comments are also made in Chapter 2 concerning the absence of positive responses to requests of the Intercenter Photo Working Group and the Debris Assessment Team for the Dept. of Defense to photograph the orbiter's underside.]

Shuttle Program Management's Low Level of Concern

The opinions of shuttle program managers and debris and photo analysts on the potential severity of the debris strike diverged early in the mission and continued to diverge as the mission progressed, making it increasingly difficult for the Debris Assessment Team to have [its] concerns heard by those in a decision-making capacity. In the face of mission managers' low level of concern and desire to get on with the mission, Debris Assessment Team members had to prove unequivocally that a safety-of-flight issue existed before shuttle program management would move to obtain images of the left wing. The engineers found themselves in the unusual position of having to prove that the situation was unsafe—a reversal of the usual requirement to prove that a situation is safe (CAIB 169).

A Lack of Clear Communication

Communication did not flow effectively up to or down from program managers. As it became clear during the mission that managers were not as concerned as others about the danger of the foam strike, the ability of engineers to challenge those beliefs greatly diminished. Managers' tendency to accept opinions that agree with their own dams the flow of effective communications.

After the accident, program managers stated privately and publicly that if engineers had a safety concern, they were obligated to communicate their concerns to management. Managers did not seem to understand that as leaders they had a corresponding and perhaps greater obligation to create viable routes for the engineering community to express their views and receive information. This barrier to communications not only blocked the flow of information to managers, but it also prevented the downstream flow of information from managers to engineers, leaving Debris Assessment Team members no basis for understanding the reasoning behind Mission Management Team decisions (CAIB 169).

The Failure of Safety's Role

Safety personnel were present but passive and did not serve as a channel for the voicing of concerns or dissenting views. Safety representatives attended meetings of the Debris Assessment Team, Mission Evaluation Room and Mission Management Team, but were merely party to the analysis process and conclusions instead of an independent source of questions and challenges. Safety contractors in the Mission Evaluation Room were only marginally aware of the debris strike analysis (CAIB 170).

Summary

Management decisions made during *Columbia*'s final flight reflect missed opportunities, blocked or ineffective communications channels, flawed analysis and ineffective leadership. Perhaps most striking is the fact that management—including shuttle program, mission management team, Mission Evaluation Room, and flight director and Mission Control—displayed no interest in understanding a problem and its implications. Because managers failed to avail themselves of the wide range of expertise and opinion necessary to achieve the best answer to the debris strike question . . . some space shuttle program managers failed to fulfill the implicit contract to do whatever is possible to ensure the safety of the crew (CAIB 170).

Chapter 7: The Accident's Organizational Causes
Organizational Cause Statement

The organizational causes of this accident are rooted in the space shuttle program's history and culture, including the original compromises that were required to gain approval for the shuttle program, subsequent years of resource constraints, fluctuating priorities, schedule pressures, mischaracterizations of the shuttle as operational rather than developmental and lack of an agreed national vision. Cultural traits and organizational practices detrimental to safety and reliability were allowed to develop, including:

• reliance on past success as a substitute for sound engineering practices (such as testing to understand why systems were not performing in accordance with requirements/specifications);

• organizational barriers which prevented effective communication of critical safety information and stifled professional differences of opinion;

• lack of integrated management across program elements;

• the evolution of an informal chain of command and decision-making processes that operated outside the organization's rules (CAIB 177).

Understanding Causes

In the board's view, NASA's organizational culture and structure had as much to do with this accident as the external tank foam.

Given that today's risks in human space flight are as high and the safety margins as razor-thin as they have ever been, there is little room for overconfidence. Yet the attitudes and decision making of shuttle program managers and engineers during the events leading up to this accident were clearly overconfident and often bureaucratic in nature. They deferred to layered and cumbersome regulations rather than the fundamentals of safety.

As the board investigated the *Columbia* accident, it expected to find a vigorous safety organization, process and culture at NASA, bearing little resemblance to what the Rogers Commission identified. NASA's initial briefings to the board on its safety programs espoused a risk-averse philosophy that empowered any employee to stop an operation at the mere glimmer of a problem. Unfortunately, NASA's views of its safety culture in those briefings did not reflect reality (CAIB 177).

The silence of program-level safety processes undermined oversight; when they did not speak up, safety personnel could not fulfill their stated mission to provide "checks and balances." A pattern of acceptance prevailed throughout the organization that tolerated foam problems without sufficient engineering justification for doing so (CAIB 178).

Chapter 8: History As Cause: *Columbia* and *Challenger*
Echoes of *Challenger*

The constraints under which the agency has operated throughout the shuttle program have contributed to both shuttle accidents. Although NASA leaders have played an important role, these constraints were not entirely of NASA's own making. The White House and Congress must recognize the role of their decisions in this accident and take responsibility for safety in the future (CAIB 195-196).

Failures of Foresight: Two Decision Histories and the Normalization of Deviance

NASA documents show how official classifications of risk were downgraded over time. Program managers designated the foam problem and the O-ring erosion as "acceptable risks" in flight readiness reviews (CAIB 196).

System Effects: The Effect of History and Politics on Risky Work

The board found that dangerous aspects of NASA's 1986 culture, identified by the Rogers Commission, remained unchanged (CAIB 198).

Pre-*Challenger* budget shortages resulted in safety personnel cutbacks. Without clout or independence, the safety personnel who remained were ineffective. In the case of *Columbia*, the board found the same problems were reproduced and for an identical reason: When pressed for cost reduction, NASA attacked its own safety system. The faulty assumption that supported this strategy prior to *Columbia* was that a reduction in safety staff would not result in a reduction of safety because contractors would assume greater safety responsibility. Post-*Challenger* NASA still had no systematic procedure for identifying and monitoring trends (CAIB 198-199).

Organization, Culture and Unintended Consequences

At the same time that NASA leaders were emphasizing the importance of safety, their personnel cutbacks sent other signals. Streamlining and downsizing, which scarcely go unnoticed by employees, convey a message that efficiency is an important goal. The shuttle/space station partnership affected both programs. Working evenings and weekends just to meet the International Space Station Node 2 deadline sent a signal to employees that schedule is important. When paired with the "faster, better, cheaper" NASA motto of the 1990s and cuts that dramatically decreased safety personnel, efficiency becomes a strong signal and safety a weak one. This kind of doublespeak by top administrators affects people's decisions and actions without them even realizing it (CAIB 199).

History as a Cause: Two Accidents

The organizational structure and hierarchy blocked effective communication of technical problems. Signals were overlooked, people were silenced, and useful information and dissenting views on technical issues did not surface at higher levels. What was communicated to parts of the organization was that O-ring erosion and foam debris were not problems (CAIB 201).

NASA's safety system lacked resources, independence, personnel and authority to successfully apply alternative perspectives to developing problems. Overlapping roles and responsibilities across multiple safety offices also undermined the possibility of a reliable system of checks and balances (CAIB 202).

Changing NASA's Organizational System

Leaders create culture. It is their responsibility to change it. Top administrators must take responsibility for risk, failure and safety by remaining alert to the effects their decisions have on the system. Leaders are responsible for establishing the conditions that lead to their subordinates' successes or failures. The past decisions of national leaders—the White House, Congress and NASA headquarters—set the *Columbia* accident in motion by creating resource and schedule strains that compromised the principles of a high-risk technology organization (203).

Part 3: Discussion Guide

An SH&E professional will need both considerable tact and diplomacy to convince management to review the history of safety decision making in order to determine whether, over time, latent technical conditions and operating practices have accumulated which could be the causal factors for a major accident. To generate interest in such a review, the author recommends that SH&E professionals send this article up through the organizational chain.

To facilitate such a review, it would be valuable to develop an outline of subjects to be discussed. An initial outline follows. It pertains specifically and only to the content of the CAIB report and, therefore, is not complete. However, it can serve as a framework for developing a discussion outline suitable to a particular operation.

1) How does management view its safety culture? How does management's view compare with the perception of employees? Does senior management's view of the safety culture reflect reality?

2) Does a gap exist between what management says and what management does?

3) Has the staff reporting directly to the senior manager been held accountable, in reality, for a high level of safety decision making?

4) Does the organization's culture gradually accept escalating risk?

5) Does the organizational structure enhance safety decision making?

6) Do organizational barriers prevent effective communication on safety, up and down?

7) Have streamlining and downsizing conveyed a message that efficiency and being on schedule are paramount, and that safety considerations can be overlooked? Does this result in "doublespeak" by management?

8) Are technical and operational safety standards at a sufficiently high level?

9) Has it been the practice to accept performance at a lesser level than that prescribed in technical and operational standards?

10) Have known safety problems, over time, been relegated to a less-than-adequate status and, thereby, become "accepted risk"?

11) Have safety-related hardware and software become obsolete?

12) Are certain operations continued with the knowledge they are unduly hazardous?

13) Have budget constraints had a negative effect on safety decision making?

14) Has inadequate maintenance resulted in an accumulation of hazardous situations that have gone unattended? For example, is detection equipment adequate, maintained and operable; are basic repairs to structures and equipment awaiting action?

15) For the opportunity to apply early interventions, has adequate attention been paid to near-hit incidents that could, under other circumstances, result in a major accident?

16) Are SH&E personnel encouraged to be tactfully aggressive when expressing their views on hazards and risks, even though their views may differ from those held by others?

17) Has it been acceptable that accident investigation stops at the first identifiable causal factor (referred to in the *Columbia* report as "the immediate technical flaw or individual failure")? Or, are accidents investigated in depth to identify the real root-cause factors so that appropriate safety interventions can be applied?

18) Has the firm relied too heavily on outside contractors (outsourcing) to do what they cannot do effectively with respect to safety?

19) Are purchasing and contracting procedures in place at a level to ensure that hazards are not introduced to the workplace?

Responses to these questions would be evaluative. What resources might an SH&E practitioner use to determine the related best practices? These publications (all available through ASSE) are a starting point:

•*Accident Investigation Techniques: Basic Theories, Analytical Methods and Applications*, by Jeffrey S. Oakley.

•*Analyzing Safety System Effectiveness, Third Edition*, by Dan Petersen.

•*Innovations in Safety Management: Addressing Career Knowledge Needs*, by Fred A. Manuele.

•*Managing for World Class Safety*, by J.M. Stewart.

- *On The Practice of Safety, Third Edition*, by Fred A. Manuele.
- *Safety Engineering, Third Edition*, by Richard T. Boehm.

Safety management systems do not often include provisions for identifying and minimizing the potential for major accidents. It seems that there is opportunity here for the enterprising. ■

References

Columbia Accident Investigation Board (CAIB). "*Columbia* Accident Investigation Board Report, Vol. 1." Washington, DC: NASA, Aug. 2003. <http://www.nasa.gov/columbia/home/CAIB_Vol1.html>.

Norman, D.A. *The Psychology of Everyday Things*. New York: Basic Books, 1988.

Reason, J. *Managing the Risks of Organizational Accidents*. Burlington, VT: Ashgate Publishing Co., 1997.

4 RISK ASSESSMENT AND HIERARCHIES OF CONTROL
THEIR GROWING IMPORTANCE TO THE SH&E PROFESSION

ORIGINALLY PUBLISHED MAY 2005

Risk assessment provisions and hierarchies of control that outline a course of action to be taken to resolve hazard and risk situations are now included in many new and revised safety standards and guidelines. These documents reflect the views of a broad cross-section of SH&E practitioners, and it is likely that future standards will include similar provisions. Therefore, SH&E professionals need to understand risk assessment methodologies and the thought processes encompassed in hierarchies of control. This article discusses:

•recently issued standards and guidelines that require risk assessments and the use of a hierarchy of control;

•the purpose of a hierarchy of control;

•a concept in which hazard identification and analysis, risk assessment and a hierarchy of controls are joined with sound problem-solving methods to create a safety decision hierarchy;

•hazard identification and analysis, and risk assessment methods.

Risk Assessment and Hierarchy of Control Provisions

Following are several examples of standards and guidelines issued in recent years that require risk assessments and the use of a prescribed hierarchy of controls. Other relevant standards are also briefly discussed. These standards and guidelines reflect the work of many SH&E professionals who have agreed that a prescribed and sequential course of action should be undertaken to effectively resolve hazard and risk situations. Although the specifics vary in the cited documents, they all share similar characteristics.

ANSI/ASSE Z244.1-2003

In July 2003, ANSI approved Control of Hazardous Energy: Lockout/Tagout and Alternative Methods. With respect to occupational safety, Z244.1-2003 may have a broader impact than any other safety standard issued in recent years. It will affect a vast number of locations. Section 5.4, which discusses alternative methods of control, is paraphrased here:

> When lockout/tagout is not used for tasks that are routine, repetitive and integral to the production process, or traditional lockout/tagout prohibits the completion of those tasks, an alternative method of control shall be used. Control options shall follow the hierarchy of alternative control implementation shown here. Selection of an alternative control method by the user shall be based on a risk assessment of the machine, equipment or process. The hierarchical control process shall be applied in the following order of preference:
> a) eliminate the hazard through design;
> b) use engineered safeguards;
> c) use warning and alerting techniques;
> d) use administrative controls (e.g., safe work procedures, training);
> e) use PPE.

This standard will have a broad impact because it requires a risk assessment *before* an alternative risk control method is selected and also presents a specific hierarchical control methodology.

ANSI B11.TR3-2000

In ANSI's identification system, "TR" stands for technical report. TR3 is titled Risk Assessment and Risk Reduction: A Guide to Estimate, Evaluate and Reduce Risks Associated with Machine Tools (AMT). This report was published in November 2000. Section 4 presents an "overview of risk assessment and risk reduction." It proposes that risk assessment methods be used in the design and use of a machine to arrive at a tolerable risk level. The following comments are from Section 8, which discusses risk reduction.

> The risk assessment process yields a level of risk (probability of occurrence of harm and the severity of that harm). The performance and ease of use of protective measures should be appropriate to the desired degree of risk reduction. Protective measures should be applied in the hierarchical order, the major captions for which are as follows.

a) Eliminate the hazard or reduce the risk by design.
b) Apply safeguards.
c) Implement administrative controls or other protective measures.

Although this hierarchy of controls seems briefer than that in ANSI Z244.1, the detail in the standard addresses the same subjects as the lockout/tagout standard. SH&E practitioners should note the difference in perception by the developers of this TR with regard to how the hierarchy of controls should be presented. Also, as ANSI B11 series standards that pertain to the design, construction, care and use of machine tools are updated, the risk assessment and hierarchy of control provisions in TR3 are being incorporated into them.

ANSI/RIA R15.06-1999

This standard, titled Industrial Robots and Robot Systems: Safety Requirements, also includes provisions for risk assessment and the use of a hierarchy of controls. Annex A, Table A.2 presents the following hierarchy of safeguarding controls:

1) elimination or substitution;
2) engineering controls (safeguarding technology);
3) awareness means;
4) training and procedures (administrative controls);
5) PPE (Robotics Industries Assn.).

This hierarchy differs from those previously cited in that it adds the element of substituting less-hazardous methods or materials as a means of attaining a tolerable risk level.

SEMI S2-0200

Semiconductor Equipment and Materials International (SEMI) is a trade group for the semiconductor industry. An updated version of SEMI S2-0200, Environmental, Health and Safety Guideline for Semiconductor Manufacturing Equipment, was issued in February 2000. Section 6.8 and 6.8.1 read as follows.

6.8 A hazard analysis should be performed to identify and evaluate hazards. The hazard analysis should be initiated early in the design phase and updated as the design matures.

6.8.1 The hazard analysis should include consideration of:
•application or process;
•hazards associated with each task;
•anticipated failure modes;
•probability of occurrence and severity of a mishap;
•level of expertise of exposed personnel and the frequency of exposure;
•frequency and complexity of operating, servicing and maintenance tasks;
•safety critical parts (SEMI).

Practitioners engaged in hazard analysis and risk assessment have not agreed on universal definitions for the terms they use. For example, the content of 6.8.1 is described as a hazard analysis. Some may consider it to be an outline for a hazard analysis and a risk assessment.

At 6.9 in the guideline, "the order of precedence for resolving identified hazards" is given: Design to eliminate hazards. Incorporate safety devices. Provide warning devices. Provide hazard warning labels. Develop administrative procedures and training (SEMI). This order of precedence is similar to that set forth in MIL-STD-882D.

MIL-STD-882D

Much of the wording in the preceding hierarchies of control is comparable to that found in military standard system safety requirements. First issued in 1969 as MIL-STD-882, the fourth edition, issued in February 2000 is designated as MIL-STD-882D (U.S. Dept. of Defense). In system safety literature, writers trace the principles embodied in military standard system safety requirements to the work of aviation and space age personnel that commenced after World War II.

The design order of precedence for mitigating hazards as it appears in 882D is an extension of the provisions to satisfy safety requirements shown in the original version of the standard. (Precedence, as used here, means priority in order, rank or importance.) The changes made were derived from 30 years of learning experience. The original requirements were: Design for minimum hazard; safety devices; warning devices; and special procedures.

Section 4.4 in 882D, "Identification of Mishap Risk Mitigation Measures," includes "the system safety design order of precedence for mitigating identified hazards." Section 4.4 follows in its entirety.

> **4.4 Identification of mishap risk mitigation measures.** Identify potential mishap risk mitigation alternatives and the expected effectiveness of each alternative or method. Mishap risk mitigation is an iterative process that culminates when the residual mishap risk has been reduced to a level acceptable to the appropriate authority.
>
> **The System Safety Design Order of Precedence for Mitigating Hazards**
> a) Eliminate hazards through design selection. If unable to eliminate an identified hazard, reduce the associated mishap risk to an acceptable level through design selection.
>
> b) Incorporate safety devices. If unable to eliminate the hazard through design selection, reduce the mishap risk to an acceptable level through using protective safety features or devices.
>
> c) Provide warning devices. If safety devices do not adequately lower the mishap risk of the hazard, include a detection and warning system to alert personnel to the particular hazard.
>
> d) Develop procedures and training. Where it is impractical to eliminate hazards through design selection or to reduce the associated risk to an acceptable level with safety

and warning devices, incorporate special procedures and training. Procedures may include the use of personal protective equipment. For hazards assigned catastrophic or critical mishap severity categories, avoid using warning, caution or other written advisory as the only risk reduction method (U.S. Dept. of Defense).

Clearly, the hierarchies of control included in recently issued safety standards and guidelines have been much influenced by the safety design order of precedence as it evolved in the several editions of MIL-STD-882.

ANSI/AIHA Z10

American Industrial Hygiene Assn. (AIHA) is secretariat of ANSI Z10, Occupational Health and Safety Systems. The scope of this draft standard is to "develop a standard of management principles and systems to help organizations design and implement deliberate and documented approaches to continuously improve their occupational health and safety performance" (AIHA). The draft standard contains extensive provisions for risk assessment and the use of a hierarchy of controls. The Z10 Committee is currently evaluating public comment on the draft in accordance with ANSI requirements.

ANSI/PMMI B155.1-2000

The Packaging Machinery Manufacturers Institute (PMMI) is secretariat of ANSI B155.1, Standard for Packaging and Packaging-Related Converting Machinery: Safety Requirements for Construction, Care and Use. Last issued in 2000, the standard is currently under revision. A review of the latest draft indicates that provisions regarding identifying and analyzing hazards, assessing risk, applying a hierarchy of controls and reducing risk to an acceptable level over the lifecycle of the packaging machinery will be expanded (PMMI). The foreword of the draft states, "This version of the standard has been harmonized with European (EN) and international (International Organization for Standardization or ISO) standards by the introduction of hazard identification and risk assessment as the principal method for analyzing hazards to personnel and achieving a level of acceptable risk" (PMMI).

European Influence

Actions in Europe have also provided some impetus to include provisions for hazard analysis, risk assessment and a hierarchy of controls in U.S. standards. Two standards are particularly relevant:

1) ISO 12100-1, Safety of Machinery: Basic Concepts, General Principles for Design—Part 1, which requires that risk assessments be conducted for machinery going into a European workplace [ISO(a)];

2) ISO 14121/EN 1050, Safety of Machinery: Principles of Risk Assessment, which sets forth risk assessment concepts [ISO(b)].

In addition, under several directives from the European Committee for Standardization, American manufacturers that export to Europe are required to place a "CE" mark on their products to indicate that the product complies with all operable directives.

Purpose of a Hierarchy of Control

A hierarchy is any system of actions, things or persons ranked one above the other. For SH&E practitioners, a hierarchy of controls establishes the actions to be considered in an order of effectiveness to resolve unacceptable hazardous situations. Achieving an understanding of the significance and the rationale for this order is an important step in the continuing evolution of the practice of safety.

For many situations, a combination of the risk management methods included in a hierarchy of controls may be applied. However, the expectation is that sequential consideration will be given to each method in a descending order, and that reasonable attempts will be made to eliminate or reduce the hazards and their associated risks by taking the more-effective steps higher in the hierarchy before lower steps are considered. A lower step is not to be chosen until practical applications of the preceding higher levels are exhausted.

The Safety Decision Hierarchy

The following observations are shared as a reflection of the author's experience encompassing the design engineering aspects, operational aspects and post-incident aspects of the practice of safety. SH&E professionals often recommend solutions for hazard/risk situations before they have defined the reality of the problem—that is, before they identify the specifics of the hazards and assess the associated risks. Rarely are systems in place to determine whether the actions that SH&E professionals recommend achieve the intended risk reduction.

These observations led to exploration of the feasibility of encompassing a hierarchy of controls within established problem-solving techniques. The techniques presented in the several problem-solving texts reviewed have great similarity. Following is a composite of those techniques.

Problem-Solving Methodology

1) Identify the problem.
2) Analyze the problem.

3) Explore alternative solutions.

4) Select and take action.

5) Examine the effects of the action taken.

Can a hierarchy of controls be encompassed within typical problem-solving techniques? Yes, it can and should be. At least one other author has done so:

> Risk engineering techniques provide a thorough, systematic approach to evaluate and reduce occupational hazards. The risk engineering approach includes the following steps.
> 1) Define the facility and environments.
> 2) Identify the hazards.
> 3) Evaluate the risk.
> 4) Develop corrective actions and/or safety design criteria.
> 5) Verify acceptability of risk (Bass 65).

Essentially, this is an adoption of well-publicized problem-solving approaches. In a sense, the Scope and Functions of the Professional Safety Position, first issued by ASSE in 1966 and updated regularly since, also presents a problem-solving methodology:

> The major areas relating to the protection of people, property, and the environment are:
> A) Anticipate, identify and evaluate hazardous conditions and practices.
> B) Develop hazard control designs, methods, procedures and programs.
> C) Implement, administer and advise others on hazard controls and hazard control programs.
> D) Measure, audit and evaluate the effectiveness of hazard controls and hazard control programs (ASSE).

Words/phrases that stand out are anticipate, identify, evaluate, develop hazard control(s), implement hazard controls and measure the effectiveness. That is fundamental problem solving. Figure 1 presents an attempt to encompass a sound hierarchy of controls within sound problem-solving techniques. A description of this process follows.

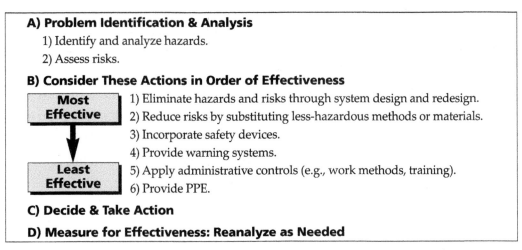

Figure 1: Safety Decision Hierarchy

Problem Identification and Analysis

Problem-solving models typically include a problem identification and analysis phase. In applying the safety decision hierarchy, the goal is to identify and analyze hazards, then assess the associated risks. Hazards and risks cannot be addressed until one determines the severity of harm that can result from a hazardous incident and assesses the probability of such an incident occurring.

Defining Risk Assessment

Unfortunately, a broadly accepted definition of risk assessment has not emerged. *Risk Assessment: Basics and Benchmarks* lists definitions from 19 sources, as well as eight definitions for risk analysis and seven for risk estimation (Main). Clearly, simplicity is needed in defining risk. The following statements are offered to help build that definition.

•Hazards are defined as the potential for harm. The dual nature of hazards must be understood. Hazards include any aspect of technology or activity that produces risk. Hazards include the characteristics of things and the actions or inactions of people.

•Risk is defined as a combination of the probability of a hazard-related incident occurring and the severity of harm or damage that could result.

•Probability is defined as the likelihood of a hazard being realized and initiating an incident or series of incidents that could result in harm or damage—for the selected unit of time, events, population, items or activity being considered.

•Severity is defined as the extent of harm or damage that could result from a hazard-related incident.

•The entirety of purpose of those accountable for safety, whatever their titles, is to manage their endeavors with respect to hazards so that their associated risks are acceptable.

Risk assessment commences with hazard identification and analysis, through which the probable severity of harm is established (assuming that a hazard's potential is realized and a hazard-related incident occurs); it concludes with an estimate of the probability of the hazard-related incident occurring. An appropriate statement indicating risk level must include both the probability of a hazard-related incident occurring (related to some statistical base) and the severity of harm or damage that could result. If a risk assessment establishes that risks are not acceptable, appropriate abatement actions would taken.

Hazard Analysis and Risk Assessment Guide

Following is a general guide on how to perform a hazard analysis and a risk assessment. Specific methods to be applied for risk assessment are not prescribed in the standards and guidelines previously cited. The intent is that the technique best-suited to the given hazard/risk situation be applied. Many such methodologies are available. For example, the *System Safety Handbook* describes 101 analytical methods (Stephans and Talso). Commonly used techniques include preliminary hazard analysis; safety reviews; operations analysis; what-if analysis; checklist analysis; what-if checklist analysis; hazard and operability analysis (HAZOP); failure modes and effects analysis; fault-tree analysis; and management oversight and risk tree [Manuele(a)].

Whatever the simplicity or complexity of the hazard/risk situation, and whatever the risk assessment methodology used, the following thought-and-action process is applicable.

1) Establish analysis parameters. Select a manageable task, system, process or product to be analyzed, and establish its boundaries and operating phase (e.g., standard operation, maintenance, startup). Determine the scope of the analysis in terms of what can be harmed or damaged: People (the public, employees), property, equipment, productivity and the environment.

2) Identify the hazards. The frame of thinking adopted should get to the bases of causal factors, which are hazards. These questions should be asked: What characteristics of things or the actions or inactions of people present a potential for harm? What aspects of the activity or technology produce risk?

Depending on the complexity of the situation, some or all of the following may apply.
- Use intuitive engineering and operational sense. This is paramount throughout.
- Examine system specifications and expectations.
- Review relevant codes, regulations and consensus standards.
- Interview current or intended system users or operators.
- Consult checklists.
- Review studies from similar systems.
- Consider the potential for unwanted energy releases and exposure to hazardous substances.
- Review historical data such as industry experience, incident investigation reports, OSHA and National Safety Council data, and manufacturers' literature.
- Brainstorm.

3) Consider the failure modes. Define possible failure modes that would result in the realization of the potentials of the hazards. Consider how an undesirable event could occur and what controls are in place to mitigate its occurrence.

4) Determine exposure frequency and duration. For each harm or damage category selected in Step 1 for the scope of the analysis, estimate the frequency and duration of exposure to the hazard (i.e., the frequency and duration of vulnerability or endangerment). For example, for workers, consider how often the task is performed, the duration of exposure and the number of people affected.

5) Assess the severity of consequences. What is the magnitude of harm or damage that could result? Learned speculations must be made regarding the consequences of an occurrence: The number of resulting injuries and illnesses or fatalities; the value of property or equipment damaged; the duration of lost productivity; or the extent of environmental damage. Historical data can establish a baseline. On a subjective basis, the goal is to determine the worst-credible consequences should an incident occur, not the worst-conceivable consequences. When the severity of consequences is determined, the hazard analysis is complete.

6) Determine occurrence probability. Consider the likelihood that a hazardous event will occur. This process is also subjective. For more-complex hazardous scenarios, it is best to brainstorm with people knowledgeable of the issues involved. Probability is to be related to an interval base of some sort, such as a unit of time or activity, events, units produced, or the life cycle of a facility, equipment, process or product.

7) Define the risk. Conclude with a statement that addresses both the probability of an incident occurring and the expected severity of harm or damage. Categorize each risk in accord with agreed-upon terms, such as high, serious, moderate or low.

8) Rank risks in priority order. Risks should be ranked in order to establish priorities. Since the hazard analysis and risk assessment exercise is subjective, the risk-ranking system will also be subjective.

9) Develop remediation proposals. When required by the results of the risk assessment, alternate proposals for design and operational changes that are needed to achieve an acceptable risk level would be recommended.

10) Take action. Action should be taken as necessary, as should follow-up activities to determine whether the action was effective.

Risk Assessment Matrixes

The author has collected 15 risk assessment matrixes—some simple, some complex. Each matrix presents categories of incident occurrence probability and the severity of harm or damage that could result. A risk assessment matrix is a method to display the combinations of probability and severity and to categorize those combinations. Such a matrix also helps the SH&E professional communicate with and influence decision makers [Manuele(a)].

It is best to adopt or develop a risk assessment matrix that is suitable to an entity's particular needs. In this process, one must ensure that the meanings of the terms contained in the matrix are understood by those making risk assessments and by decision makers. While various matrixes may contain the same terms—such as the probability of an incident occurring being frequent or likely, and the severity of consequences being high or serious—they may have different descriptions. For illustration, Table 1 presents a risk assessment matrix while Table 2 illustrates possible management decision levels.

OCCURRENCE PROBABILITY	SEVERITY OF CONSEQUENCES			
	Catastrophic	Critical	Medium	Minimal
Frequent	High	High	Serious	Moderate
Likely	High	High	Serious	Moderate
Occasional	Serious	Serious	Moderate	Low
Remote	Moderate	Moderate	Moderate	Low
Improbable	Low	Low	Low	Low

Table 1: Risk Assessment Matrix

Risk category	Remedial action or acceptance
High	Operation not permissible.
Serious	Remedial action to have high priority.
Moderate	Remedial action to be taken in appropriate time.
Low	Risk is acceptable; remedial action is discretionary.

Table 2: Management Decision Levels

The Logic of Taking Action in an Order of Effectiveness

In the safety decision hierarchy, alternative risk reduction and elimination actions are listed in descending order of effectiveness. Sound safety management requires that one establish the rationale for the order in which the list is presented.

Actions described in the first, second and third levels are more effective because they 1) are preventive actions that reduce risk by design and substitution measures; 2) rely least on personnel performance; and 3) are less defeatable. Actions described in the fourth, fifth and sixth levels rely greatly on the performance of people.

Action Level 1

If the hazards are eliminated in the design and redesign processes, risks that derive from those hazards are also eliminated. The intent is to design to acceptable risk and to minimize the human action necessary in the work process. Examples include designing to eliminate hazards related to falls, ergonomics, confined spaces, noise and chemicals. If no hazards are present, there is no potential for harm, and, thereby, no risk.

Action Level 2

By substituting less-hazardous methods or materials, risks can be substantively reduced. Examples include using automated materials handling equipment rather than manual materials handling; providing an automatic feed system to reduce machine hazards; using a less-hazardous cleaning material; and replacing an old steam heating system and its boiler explosion hazards with a hot air system. This reduces the need to rely on the actions of people, although perhaps not to the same extent as designing out the hazard.

Note that this hierarchy of controls separates eliminating hazards and risks in the design process from substituting less-hazardous methods/materials. Based on the author's experience, substitution of a less-hazardous method/material may or may not result in equivalent risk reduction in relation to what might be the case if hazards and risks are reduced through design engineering.

Consider this example: The mixing process for chemicals often requires considerable manual materials handling. A reaction occurs and an employee sustains serious chemical burns.

Identical operations are performed at two of this company's locations. At one, management decides to re-engineer the operation so it is completely enclosed, automatically fed and operated by computer from a control panel. At the other location, no funds are available for re-engineering, so site management arranges for the supplier to premix the chemicals before shipment and installs some mechanical feed equipment for the chemicals. The risk reduction achieved as a result would not be equivalent to that attained by re-engineering the operation.

In another example, if a 110-volt power source replaces an 880-volt power source, the injurious power level has been reduced, but 110 volts with the necessary amperage can still be fatal.

Action Level 3

When safety devices are incorporated into the system or product in the form of engineering controls, risk can be reduced, as can reliance on the worker or product user's actions. Safety devices include machine guarding, interlock systems, presence-sensing devices, safety nets, fall prevention systems, and all devices and systems that separate hazardous energy from personnel.

Action Level 4

Warning systems, although vital in many situations, are reactionary. They alert people only after a hazard's potential is in the process of being realized (e.g., a smoke alarm). Warning system effectiveness and the effectiveness of instructions, signs and warning labels rely considerably on administrative controls, training, the quality of maintenance and people's reactions.

A note about the term "warning systems." In one published hierarchy of control, the designation for this purpose is merely "warning signs"; in another, it is simply "warnings." The entirety of the needs of a warning system must be considered, for which warning signs alone may be inadequate. For example, NFPA Life Safety Code 101 may require, among other things: smoke and products-of-combustion detectors; automatic and manual audible and visible alarms; lighted exit signs; designated, alternate, properly lit exit paths; adequate spacing for personnel at the end of the exit path; proper hardware for doors; and emergency power systems.

Action Level 5

Administrative controls include appropriate work methods and procedures, personnel selection, training, supervision, motivation, work scheduling, job rotation, scheduled rest periods, maintenance, management of change, investigations, inspections and behavior modification. These controls rely on the appropriateness of the particular method in relation to needs, capabilities of those responsible for their delivery and application, quality of supervision and performance of workers. It is difficult to achieve a superior level of effectiveness in all these areas.

Action Level 6

Proper use of PPE—such as safety glasses, face shields, safety shoes, gloves and hearing protection—relies on an extensive series of supervisory and personal actions, such as the identification of the equipment needed, and its selection, fitting, training, inspection and maintenance. Although use of PPE is common and it is necessary in many occupational situations, it is the least-effective method to address hazards and risks; it is also a method that can be easily defeated.

Deciding and Taking Action

The next step is to decide on and take action. Once this decision is made, some fundamental management practices are necessary, such as assigning responsibility, scheduling, providing resources (staffing and money) and setting target dates for completion—all of which must be documented.

Measuring for Effectiveness and Reanalyzing as Necessary

Ensuring that actions taken accomplish their intended goal is an integral step in an effective problem-solving technique. For safety management purposes, measuring for effectiveness requires verifying whether actions taken have truly reduced the risk to the level expected. Follow-up activity would determine whether the solutions were effective.

- The problem was resolved, only partially resolved or not affected.
- All hazards were or were not addressed.
- Actions taken did or did not create new hazards.

No matter what actions are taken, if a work activity continues, it will always produce residual risk (risk that remains after preventive measures have been taken). It is not possible to attain zero risk. If the residual risk is not acceptable, the thought process involved with the safety decision hierarchy must be reapplied, beginning with hazard identification and analysis process.

Conclusion

Requirements for risk assessment and hierarchies of control are now common components of safety standards and guidelines. As these provisions become more prevalent, SH&E professionals must takes steps to understand them in order to effectively apply these techniques. ∎

References

ASSE. Control of Hazardous Energy-Lockout/Tagout and Alternative Methods. ANSI/ASSE Z244.1-2003. Des Plaines, IL: ASSE, 2003.

ASSE. "Scope and Functions of the Professional Safety Position." Des Plaines, IL: ASSE.

Assn. for Manufacturing Technology (AMT). "Risk Assessment and Risk Reduction: A Guide to Estimate, Evaluate and Reduce Risks Associated with Machine Tools." ANSI B11.TR3 2000. McLean, Virginia: AMT, 2000.

American Industrial Hygiene Assn. (AIHA). Draft American National Standard on Occupational Health and Safety Systems. ANSI/AIHA Z10. Fairfax, VA: AIHA.

Bass, L. "Risk Engineering: Keys to Applying Techniques to Occupational Health & Safety." *Industrial Safety and Hygiene News.* June 2004: 65-66.

Bransford, J.D. and B.S. Stein. *The Ideal Problem Solver.* New York: W.H. Freeman and Co., 1984.

European Committee for Standardization. General Product Safety Directive. 2001/95/EC. Brussels: European Committee.

International Organization for Standardization (ISO)(a). Safety of Machinery: Basic Concepts, General Principles for Design—Part 1. ISO 12100. Geneva, Switzerland: ISO, 2003.

ISO(b). Safety of Machinery: Principles of Risk Assessment. ISO 14121/EN 1050. Geneva, Switzerland: ISO, 1999.

Kepner, C.H. and B.B. Tregoe(a). *The Rational Manager.* New York: McGraw-Hill, 1965.

Kepner, C.H. and B.B. Tregoe(b). *The New Rational Manager.* Princeton, NJ: Princeton Research Press, 1981.

Main, B.W. *Risk Assessment: Basics and Benchmarks.* Ann Arbor, MI: Design Safety Engineering Inc., 2004.

Manuele, F.A.(a). *Innovations in Safety Management: Addressing Career Knowledge Needs.* Hoboken, NJ: John Wiley & Sons, 2001.

Manuele, F.A.(b). *On The Practice of Safety.* 3rd ed. Hoboken, NJ: John Wiley & Sons, 2003.

Packaging Machinery Manufacturers Institute (PMMI). Standard for Packaging and Packaging-Related Converting Machinery: Safety Requirements for Construction, Care and Use. ANSI/PMMI B155.1. Arlington, VA: PMMI, 2000.

Robotics Industries Assn. American National Standard for Industrial Robots and Robot Systems: Safety Requirements. ANSI/

RIA R15.06-1999. Ann Arbor, MI: Robotic Industries Assn., 1999.

Semiconductor Equipment and Materials International (SEMI). Environmental, Health and Safety Guidelines for Semiconductor Manufacturing Equipment. SEMI S2-0200. Mountain View, CA: SEMI, 2000.

Stephans, R.A. and W.W. Talso. *System Safety Analysis Handbook.* 2nd ed. Albuquerque, NM: System Safety Society, New Mexico Chapter, 1997.

U.S. Dept. of Defense. Military Standard System Safety Program Requirements. MIL-STD-882D. Washington, DC: U.S. Dept. of Defense, 2000.

5 THE CHALLENGE OF PREVENTING SERIOUS INJURIES
A PROPOSAL FOR SH&E PROFESSIONALS

ORIGINALLY PUBLISHED APRIL 2006

The frequency of worker injuries is down, but serious injuries are more prominent within the entirety of the lost-worktime cases reported and average workers' compensation claims costs have risen at a remarkable rate. This trend suggests the need for SH&E professionals to study the characteristics of incidents resulting in serious injury, particularly of the nature of work being performed and the job titles of injured personnel.

In *The Blame Machine: Why Human Error Causes Accidents*, Whittingham describes how disasters and serious accidents result from recurring but potentially avoidable human errors. He shows how such errors are preventable because they result from defective systems within a company. Based on analyses of several events, he identifies the common causes of human error and the typical system deficiencies that led to those errors. Those deficiencies were principally organizational, cultural, technical and management systems failures.

According to Whittingham, a "blame culture" exists in some organizations, whereby the focus of investigations of incidents that result in severe consequences is on individual human error, and the focus of corrective action taken is at that level rather than on the system that may have enabled the human error.

Early on, Whittingham says that many organizations—and sometimes entire industries—are unwilling to look closely into error-provocative system faults. He stresses that putting responsibility for the incident on what an individual did/did not do results in simplistic causal factor determination (Whittingham). Many SH&E professionals should think about this when dealing with their clients.

In organizations reluctant to explore systemic causal factors, the incident investigation stops after addressing the individual human error—the unsafe act. Thus, a more-thorough investigation that looks into the true root-causal factors is avoided. If an organization chooses to reduce the probability of serious injuries, safety management systems must be in place to:

•anticipate and take corrective action on hazards that may have serious injury potential;

•ensure in-depth reviews of root-causal factors for incidents that result in serious injury;

•address organizational, operational, technical and cultural causal factors.

As practitioners study serious injury trending, they may find that a culture change is necessary to achieve desired goals.

Statistical Indicators

Data displaying the adverse progression for serious injuries and workers' compensation claims costs have been extracted from two sources—Bureau of Labor Statistics (BLS) and the National Council on Compensation Insurance (NCCI). It should be noted that the impact of serious injury trending over the last several years may differ considerably by industry. Statistics given here are derived from macro studies or may relate to specific industries. Practitioners should make their own studies of serious injury trending in the entities to which they provide counsel so that the conclusions they draw and the recommendations they make have a sound, realistic base.

Bureau of Labor Statistics

For many years, BLS has issued reports titled "Lost-Worktime Injuries and Illnesses: Characteristics and Resulting Time Away from Work." Data in Tables 1 and 2 and Figure 1 were taken from Table 10 (Percent Distribution of Nonfatal Occupational Injuries and Illnesses Involving Days Away from Work) in those reports for the years 1995 through 2001. Table 10 shows the percentage of selected days-away-from-work (DAFW) categories as each category relates to the total number of DAFW cases reported in a given year (BLS).

Chapter 5: The Challenge of Preventing Serious Injuries

	1	2	3-5	6-10	11-20	21-30	31 or more
1995	16.9	13.4	20.9	13.4	11.3	6.2	17.9
2001	15.4	12.7	19.8	12.6	11.1	6.3	22.0
Change from 1995	-8.9	-5.2	-5.3	-6.0	-1.8	+1.6	+22.9

Table 1: Percentage of DAFW Cases, Numbers of Days, Private Industry (BLS)

Year	Percent
1995	17.9
1996	18.5
1997	18.5
1998	19.1
1999	19.6
2000	21.0
2001	22.0
2002	23.4 (projected)
2003	25.0 (projected)

Table 2: Percentage of DAFW Cases, 31 or more DAFW

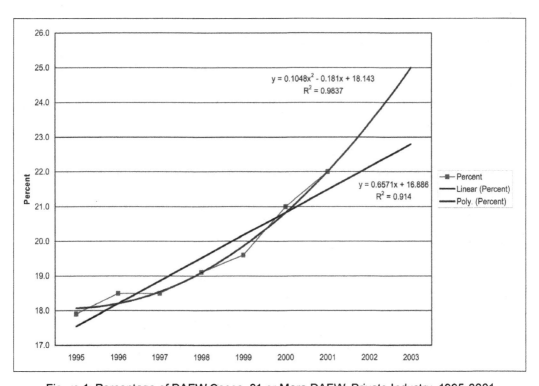

Figure 1: Percentage of DAFW Cases, 31 or More DAFW, Private Industry, 1995-2001

From 1995 to 2001, several observations can be made with respect to the changes in the distribution of the total number of lost workday cases reported in each of those years:

• The decreases in the percentages for the first four DAFW categories listed are significant.

• For the 11-to-20 and the 21-to-30 DAFW categories, the decrease of 1.8% and the increase of 1.6% are not as significant.

• The 23% increase for the 31-or-more DAFW category is significant.

Displaying the Trending

To produce more specific trend data on the lost workday cases with 31 or more DAFW, several indicators were extracted from BLS reports. Table 2 shows percentage of cases involving 31 or more DAFW from 1995 to 2001, and projected for 2002 and 2003. Note that these projections are based on the statistical history for the years 1995 through 2001 and assume that the trending in those years will continue. Rules for reporting lost workday cases were revised for 2002 with respect to how DAFW are counted. Direct comparisons for that year and subsequent years cannot be made with data for previous years.

Alan Hoskin, manager of the statistics department at National Safety Council (NSC), agrees that this trending is statistically meaningful. Using the base data for the years 1995 to 2001, he produced the projected numbers shown in Table 2 through polynomial charting. (Polynomial: A mathematical expression consisting of the sum of a number of terms, each of which contains a constant and variables raised to a positive integral power.)

For 2002, the polynomial projection for lost workday cases with 31 or more DAFW is 23.4%. In the BLS report for the year 2002 (the last year for which data are available), the recording is 25.1%. Some of this difference could result from the change in reporting rules on how DAFW are counted. For 2003, the projection is 25.0% on the polynomial line.

Figure 1 displays the trending from 1995 to 2001 and a projection of the trending into 2002 and 2003. SH&E practitioners should consider what this display means for the operations with which they are involved.

Taken as a whole, BLS data on lost-worktime injuries and illnesses seem to track well with the statements made by NCCI indicating that the frequency of workers' compensation claims has declined, with the decline being most pronounced for smaller claims.

National Council on Compensation Insurance

According to a Jan. 7, 2004, NCCI news bulletin:

> On-the-job claim frequency for workers' compensation injuries continues to decline. Research shows that this decline has been most pronounced for smaller claims. However, medical and indemnity costs continue to rise. NCCI has reported significant increases in both the average indemnity and average medical cost of a claim over the last several years.
>
> Workers' compensation indemnity claims [increased] an average of 7.4% annually since 1996. Medical claim cost trends show double-digit increases the last two years. In 2001, medical severity increased by 12%, and in 2002, medical costs rose an additional 11% [NCCI(b)].

Mealy authored an extensive report for NCCI titled "State of the Line: Analysis of Workers' Compensation Results" in May 2004. Of several exhibits in that report, two are relevant here—Exhibit 11: The Rate of Change in Workers' Compensation Indemnity Claim Costs Has Accelerated Since 1995; and Exhibit 12: Workers' Compensation Medical Cost Trends Continue to Climb (Mealy).

Data in those two exhibits are for lost-time claims. The following computations were made for the years 1996 through 2002, with 1995 having a base of one: From January 1996 through December 2002, the average indemnity claim cost increased 66% while the medical claim cost increased 83%.

To determine how those cost increases related to economic inflation as a whole, the author used an inflation data calculator found at www.inflationdata.com to determine that (with 1995 having a base of one) the accumulated inflation for the U.S. economy from January 1996 through December 2002 was 17.2%. In those years, the increases in average workers' compensation indemnity and medical claim costs were at a factor of about 4.4 times inflation. That is significant and deserves attention—particularly if the trend continues.

Consider also the following excerpts from a Feb. 3, 2005, NCCI bulletin:

> Accident year 2003 frequency decreased about 4% from the level of 2002. This is the same rate of decline as the average over the prior four years. Indemnity severity for accident year 2003 increased by about 5%.
>
> The average annual increase for the years 1999 to 2002 is approximately 8%. Medical severity, however, continued to increase at a historically high rate. While the previous four years averaged 10% annual increases, accident year 2003 increased by 11% [NCCI(a)].

To summarize, overall, the frequency of worker injuries is down; serious injuries are more prominent within the entirety of the lost-worktime cases reported; and average workers' compensation claims costs have risen at a remarkable rate.

Actions to Be Taken

In light of recent developments, SH&E practitioners need to determine precisely how these data might apply to the operations to which they give counsel. It is likely that the increases in serious cases are not identical across all industries.

As a first step, SH&E practitioners should become aware of the content of the BLS's annual publication, "Lost-Worktime Injuries and Illness: Characteristics and Resulting Days Away from Work." It contains 15 tables that provide a great variety of data on the characteristics of lost-worktime injuries and illnesses—such as for occupations of the injured persons, nature of injury and illness, experience by industry, and events or exposures from which the injuries and illnesses derive.

Practitioners should also review incidents resulting in severe consequences in their entities so they can identify the types of activity that produce these incidents and the job titles of those involved.

Basis for This Proposal: Investigation Review

Based on a review of more than 1,000 incident investigation reports, the author's analyses reveal several key findings:

•A large proportion of incidents resulting in severe injury occur in unusual and nonroutine work, in nonproduction activities and where sources of high energy are present. They also occur in what can be called at-plant construction operations. (An example of at-plant construction: A motor that weighs 800 pounds and sits on a platform 15 feet above the floor needs to be replaced; the work will be performed by in-house personnel.)

•Causal factors for low-probability/high-consequence events are seldom represented in the analytical data on accidents that occur frequently (although some ergonomics-related incidents are the exception).

•Many incidents resulting in serious injury are unique and singular events, having multiple, complex causal factors that may have technical, operational systems, or cultural origins.

Petersen has made similar observations. He supports the view that severe injury potential needs special attention:

> If we study any mass data, we can readily see that the types of accidents that result in temporary total disabilities are different from the types of accidents resulting in permanent partial disabilities or in permanent total disabilities or fatalities.
>
> The causes are different. There are different sets of circumstances surrounding severity. Thus, if we want to control serious injuries, we should try to predict where they will happen. Studies in recent years suggest that severe injuries are fairly predictable in certain situations. Some of those situations involve:

- unusual, nonroutine work;
- nonproduction activities;
- sources of high energy;
- certain construction situations (Petersen).

As noted, the causes and circumstances surrounding severity are different. SH&E practitioners should try to predict where serious injuries may occur since the occurrence of severe injuries is fairly predictable in certain types of work.

Since the characteristics of serious injuries and the types of activities and exposures in which many of them occur have been known for some time, one could question why they have not received more attention in the safety community. Two key barriers are addressed later in this article; they are related to these beliefs that have long permeated the practice of safety:

- Reducing incident frequency will equivalently reduce the occurrence of low probability/serious consequence events.
- Unsafe acts of workers are the principal causal factors for occupational incidents.

Support for the Proposal to Study Severe Injuries

A statistical history supports proposing that SH&E practitioners study the characteristics of incidents resulting in serious injury, particularly the nature of work being performed and the job titles of injured personnel.

UAW Data

At a workshop held in April 2004, Franklin Mirer, director of the United Auto Workers (UAW) health and safety department, said that over a period of 20 years, skilled trades personnel—about 20% of the UAW membership of about 700,000—had 41% of the fatalities. Skilled tradespeople are maintenance personnel, millwrights, tinsmiths, machinists, electricians and steamfitters; they are not production workers. Mostly, they perform nonroutine work, are exposed to sources of high energy and may perform at-plant construction. Hours worked during the period Mirer references are in the billions. These fatality numbers are statistically significant.

General Motors

In an article about culture change at General Motors, Simon and Frazee state:

> Statistics showed that 80% of all serious accidents at GM occurred among the skilled trades, not on the assembly line (Simon and Frazee). [According to one of the authors, for the article, serious meant life threatening.]

Liberty Mutual's 2003 Workplace Safety Index reads:

> A small percentage of workers' compensation claims continue to be responsible for the bulk of direct costs. In 2000, disabling workplace injuries were 18% of workers' compensation claims but 93% of direct costs (Liberty Mutual).

An article issued some 25 to 30 years ago by Employers Insurance Company of Wausau included data similar to that in the Liberty Mutual Index.

> A study showed that 86% of total injuries produced only 6% of total costs, while 14% of total injuries produced 94% of total costs.
> Here we can distinguish between the "trivial many" and the "vital few' (Employers of Wausau).

The report concludes, "It becomes readily apparent that the logical approach to effective loss control is to concentrate major efforts on the 'vital few.'"

That a small percentage of workers' compensation claims represent a very large proportion of total costs fits well with Pareto's Law, which is commonly referred to as the 20/80 rule and the law of the trivial many and the critical few. In a large statistical sampling, 20% of the units will represent 80% of the impact, as well as the opportunity for improvement. Spending a disproportionate amount of time on the 80%, the abundant literature on Pareto's Law shows, may achieve little return. Giving additional emphasis to the critical few is the theme of this article.

Another Study: Principal Business Operations Personnel vs. Ancillary Personnel

In February 2004, the author conducted a study to determine what percentage of lost workday cases with DAFW occurred to personnel engaged in a company's principal business operation (making a product or providing a service) and what percentage occurred to ancillary or support personnel.

The sample was small and the variations by company were considerable. Participants also provided OSHA incidence and lost workday case rates. Some of the companies with high OSHA rates had higher percentages of lost workday cases with DAFW involving workers engaged in the "principal business" than for ancillary workers. The opposite was true for companies where the OSHA rates were low for their industry classes.

Consider the possible significance of the following. The three largest companies that provided data had a total of 230,000 employees. Each company had an OSHA recordable rate below 0.5 and a lost workday case rate of less than 0.2. A composite of the data for those companies indicated that 74% of lost workday cases with DAFW occurred to ancillary and support personnel rather than to employees engaged in the principal business. Two safety directors asked to contribute data for the study said that the study was unnecessary since they believed that if incident frequency was reduced, severity potential would also be comparably reduced. That and another burdening premise need exploration.

Barriers to the Prevention of Serious Injuries

As noted, two beliefs in safety have long served as barriers to making the necessary inquiry into the reality of design and engineering, operational systems and cultural causal factors for incidents that result in serious injury. Those beliefs are:

1) Reducing incident frequency will equivalently reduce the occurrence of low probability/serious consequence events.

2) Unsafe acts of workers are the principal causal factors for occupational incidents.

In a speech at the 2003 Behavioral Safety Now Conference, Liberty Mutual's James Johnson stated:

> I'm sure that have many of us have said at one time or another that frequency reduction will result in severity reduction. This popularly held belief is not necessarily true. If we do nothing different than we are doing today, these types of trends will continue (J. Johnson).

Consider those words. If SH&E professionals and their employers do nothing different than they are doing today, the adverse trending for serious injuries will continue. Frequency reduction does not necessarily produce equivalent severity reduction. Those statements are supported by statistics.

NSC Data

The following is extrapolated from *Injury Facts*, 2003 Edition.

> From 1973 to 2001, the occupational injury and illness rate for private industry dropped 50%—from 11.3 to 5.7. In the same period, the incidence rate for total lost workday cases decreased 18%—from 3.4 to 2.8 [NSC(b)].

Obviously, the reduction in the lost workday incidence rate did not equal the reduction in incident frequency. These data on injury trending are important in that they contravene the commonly held belief that efforts concentrated on reducing injury frequency will equivalently affect injury severity.

To hold with that belief, one must assume that the causal factors for incidents which occur frequently are the same as those for incidents that result in serious injury. This author's studies show that the causal factors for incidents resulting in severity are to a large extent different and that they are rarely found in the causal data on incidents which occur frequently.

The premise that workers' unsafe acts are the principal causal factors for occupational injuries must be addressed as well. Unfortunately, many SH&E practitioners have promoted safety management systems that focus extensively on what the worker does—meaning on the prevention of unsafe acts. Furthermore, some management personnel have been taught by those practitioners that the focus of their safety management systems should be principally on worker behavior.

Adopting that mindset results in the allocation of resources predominantly to the worker behavioral aspects of safety. This results in inadequate attention on systemic causal factors deriving from design and engineering shortcomings, the hazards in current operational procedures and the system of expected behavior that has developed over time.

Greater progress in the prevention of incidents resulting in serious injury will not occur as long as these two premises remain as barriers to determining the true causal factors of incidents.

Specifics for the Study Proposed

A study of incidents resulting in serious injury will not be time consuming. The data to be collected and analyzed should already exist or be easy to obtain. The following action outline can be modified to fit particular needs.

•Define the parameters for the incidents to be studied. For example, lost workday cases involving 21 or more days away from work; or lost workday cases involving 31 or more days away from work; or cases valued at $25,000 or more; or cases valued at $50,000 or more.

•Gather incident investigation and injury data related to the serious injury definition chosen for at least a three-year period.

•For each incident:

   1) Record the nature of the work being performed.

   2) Note the job titles of the injured personnel.

   3) Determine whether the injured persons were engaged in the entity's principal business or whether they were ancillary personnel.

   4) Identify the causal factors (e.g., design and engineering, operational system, cultural, organizational).

•Analyze and summarize the data to determine what modifications in safety management systems should be proposed.

If the money value of injuries and illnesses is to be used in selecting a severity category, the following information may be helpful. A major third-party administrator analyzed 280,000 workers' compensation claims that it managed in 2003. These are some of the results from a yet-to-be-published article based on that analysis:

•Three percent of claims were valued at $25,000 to $50,000; they represented 20% of total claims costs.

•Three percent of claims were valued over $50,000; they represented 52% of total claims costs.

•Six percent of claims valued at $25,000 or more produced 72% of total claims costs.

Incident Investigation

While it is suggested that an attempt be made to identify all causal factors for the incidents to be analyzed, SH&E practitioners should not be surprised to find incident investigation reports lacking in-depth causal factor determination.

As noted, the author has studied more than 1,000 incident investigation reports provided by corporate safety directors in large companies. The purpose was to assess the quality of the investigation systems in place. On a scale of 10, with 10 being best, some companies scored a 2. Causal factor determination was dismal, meaning that opportunities to readjust the focus of preventive efforts to the benefit of workers and employers were lost.

The *Columbia* Accident Investigation Board (CAIB) drew similar conclusions during its investigation of the *Columbia* space shuttle explosion. SH&E professionals should consider how these excerpts from CAIB's report relate to the incident investigation systems of the entities with which they are involved.

> Many accident investigations do not go far enough. They identify the technical cause of the accident, and then connect it to a variant of "operator error." But this is seldom the entire issue.
>
> When the determinations of the causal chain are limited to the technical flaw and individual failure, typically the actions taken to prevent a similar event in the future are also limited: fix the technical problem and replace or retrain the individual responsible. Putting these corrections in place leads to another mistake—the belief that the problem is solved.
>
> Too often, accident investigations blame a failure only on the last step in a complex process, when a more comprehensive understanding of that process could reveal that earlier steps might be equally or even more culpable.
>
> In this Board's opinion, unless the technical, organizational and cultural recommendations made in this report are implemented, little will have been accomplished to lessen the chance that another accident will follow (CAIB).

Reason and Hobbs offer similar comments about the treatment of human error.

> Errors are consequences not just causes. They are shaped by local circumstances: by the task, the tools and equipment and the workplace in general.

> If we are to understand the significance of these factors, we have to stand back from what went on in the error maker's head and consider the nature of the system as a whole (Reason and Hobbs).

Reason and Hobbs emphasize looking into the system as a whole to identify causal factors. As a result of the analyses made of causal factors for incidents which result in serious injury, this author suggests that incident investigation should be considered as:

•A prime source for selecting leading indicators for safety management system improvement. If incident investigation is done well, the reality of the technical, organizational, methods of operation and cultural causal factors in the work system will be revealed.

•Deserving of a much higher place within all of the elements of a safety management system. The quality of incident investigation emerges as being one of the primary markers in evaluating an organization's safety culture.

To Reduce Serious Injury Potential

In addition to the improvements in safety management systems that may become apparent from the proposed studies, SH&E practitioners should take proactive measures and consider initiating the following actions.

1) Conduct a needs assessment to determine how much creative destruction and educational reconstruction is needed to: a) counter the beliefs that focusing on reducing incident frequency will equivalently reduce severity, and unsafe acts of workers are the principal causal factors; and b) recognize the benefits to be gained as a result of an additional emphasis on serious injury prevention.

If it is believed that concentrating on incidents which occur frequently encompass severity potential and that the unsafe acts of employees are the principal causes of accidents, a major culture change will be necessary to reduce the potential for low-probability/severe-consequence incidents.

In any case, educational reconstruction may be needed to achieve a mindset that recognizes the particulars of severe injury potential and the opportunities that such recognition presents. This newly adopted mindset should affect every element in the safety management system so as to cause all involved to continuously think about identifying severe injury and illness potentials, and reducing the probability of the potentials being realized, with an emphasis on being proactive and anticipatory. That will be particularly significant in the design processes that affect facilities, equipment and work methods.

In conducting a needs assessment, it should be understood that traditional safety management systems do not include activities to anticipate and identify the causal factors for low-probability/severe-consequence accidents—nor do they include particularly crafted efforts for their prevention.

Reason notes that occupational safety approaches directed largely on the unsafe acts of persons have limited value with respect to the insidious accumulation of latent practices and conditions which are typically present when organizational accidents occur (Reason).

2) Promote the adoption of a pre-job planning and safety analysis system for unusual and nonroutine work, since a large percentage of incidents that result in serious injury occur in that type of activity. This idea is gaining momentum. For example, two large companies have recently set goals to have pre-job planning systems in place for nonroutine work in all of their operations. Pre-job planning systems help to improve both productivity and risk reduction [Manuele(e)].

3) Encourage institution of a variation of the critical incident technique to gather worker comments on the hazards and risks that present serious injury potential, especially concerning near-hit incidents which could have had serious consequences under slightly different circumstances. Adopting such a system can improve upward communications from workers if their knowledge and skills are properly respected. Valuable information can be obtained from the application of this relatively inexpensive data-gathering method.

In a bulletin on ergonomics methods and tools titled "Task Analysis Methods: Critical Incident Technique," the authors state, "The critical incident technique is inexpensive and provides rich information. This technique is helpful in emphasizing the features that will make a system particularly vulnerable" (Infopolis).

Speaking of incident recall, Johnson observes:

> Incident recall is an information-gathering technique whereby employees (participants) describe situations they have personally witnessed involving good and bad practices and safe and unsafe conditions. Such studies, whether by interview or questionnaire, have a proven capacity to generate a greater quantity of relevant, useful reports than other monitoring techniques, so much so as to suggest that their presence is an indispensable criterion of an excellent safety program (W.G. Johnson).

A system that seeks to identify causal factors before their potentials are realized would serve well in attempting to avoid low-probability/severe-consequence events. Such a system can be installed inexpensively [NSC(a); Manuele(d); Tarrants; Welker, et al; Infopolis].

4) Assess the quality of incident investigations and arrange for improvement so that investigations address true causal factors and so that the reports can be a source

for selecting leading indicators for serious injury prevention. As noted, a culture change may be needed to achieve this.

Who Benefits?

Who benefits from an extended focus on the prevention of incidents that result in serious injury? Few entities have all the resources, staffing and funds needed to address every hazard and every risk deriving from those hazards. It is simply reality that some things pertaining to safety do not get done.

Therefore, priorities should be set so that the available resources are applied to do the most good. If a safety management system includes specific provisions for the prevention of severe-consequence incidents and they are successful:

- workers benefit since they will have fewer serious injuries;
- employers benefit because the escalating workers' compensation costs and their relative indirect costs would be substantially reduced.

Conclusion

The goal of this article is to encourage SH&E practitioners to identify the characteristics of the "critical few" incidents that result in serious injury and to take action to minimize the potential for their occurrence. While the adverse trending of incidents resulting in serious injuries has negative implications, it also provides opportunities.

As noted, achieving greater effectiveness in reducing serious potential may require changes in an organization's safety culture, which is not easy to do. However, SH&E professionals are obligated to perform the analyses necessary to identify the safety management system modifications needed to achieve a more effective focus on serious injury prevention. If that is achieved, the beneficial results can be substantial. ■

References

Bureau of Labor Statistics (BLS). "Lost-Worktime Injuries and Illnesses: Characteristics and Resulting Time Away from Work." Washington, DC: U.S. Dept. of Labor, BLS. <http://www.bls.gov/iif>.

Columbia Accident Investigation Board (CAIB). *Columbia Accident Investigation Report, Vol. 1.* Washington, DC: CAIB, August 2003. <http://www.nasa.gov/columbia/home/CAIB_Vol1.html>.

Employers of Wausau. "Pareto's Law and the Vital Few." Wausau, WI: Employers Insurance Company of Wausau, undated.

Infopolis 2 Consortium. "Task Analysis Methods: Critical Incident Technique." Aix-en-Provence, France: Infopolis 2 Consortium. <http://www.ul.ie/~infopolis/methods/incident.html>.

Johnson, J. Speech at 2003 Behavioral Safety Now Conference, Reno, NV.

Johnson, W.G. *MORT Safety Assurance Systems*. New York: Marcel Dekker, 1980.

Liberty Mutual Insurance Co. "2003 Liberty Mutual Workplace Safety Index." Boston: Liberty Mutual Insurance Co., 2003.

Manuele, F.A.(a). "Injury Ratios." *Professional Safety*. Feb. 2004: 22-30.

Manuele, F.A.(b). *Innovations in Safety Management: Addressing Career Knowledge Needs*. New York: John Wiley & Sons, 2001.

Manuele, F.A.(c). "Is a Major Accident about to Occur in Your Operations?" *Professional Safety*. May 2004: 22-28.

Manuele, F.A.(d). *On the Practice of Safety*. 3rd ed. New York: John Wiley & Sons, 2003.

Manuele, F.A.(e). "Severe Injury Potential: Addressing an Often-Overlooked Management Element." *Professional Safety*. Feb. 2003: 26-31.

Mealy, D. "State of the Line: Analysis of Workers' Compensation Results." Boca Raton, FL: NCCI, May 2004.

National Council on Compensation Insurance (NCCI)(a). "Countrywide Frequency and Severity." NCCI Insurance News Bulletin. Boca Raton, FL: NCCI, Feb, 3, 2005.

NCCI(b). "NCCI Reports Mixed News in Claim Frequency Numbers." NCCI Insurance News Bulletin. Boca Raton, FL: NCCI, Jan. 7, 2004.

National Safety Council (NSC)(a). *Accident Prevention Manual: Administration and Programs*. 12th ed. Itasca, IL: NSC, 2001.

NSC(b). *Injury Facts*. 2003 ed. Itasca, IL: National Safety Council, 2003.

Petersen, D. *Safety Management*. 2nd ed. Des Plaines, IL: ASSE, 1998.

Reason, J. *Managing the Risks of Organizational Accidents*. Burlington, VT: Ashgate Publishing Co., 1997.

Reason, J. and A. Hobbs. *Managing Maintenance Error: A Practical Guide*. Burlington, VT: Ashgate Publishing Co., 2003.

Simon, S.I. and P.R. Frazee. "Building a Better Safety Vehicle: Leadership-Driven Culture Change at General Motors." *Professional Safety*. Jan. 2005: 36-44.

Tarrants, W. *The Measurement of Safety Performance*. New York: Garland Press Publishing, 1980.

Welker, P.A., et al. *The Critical Incident Technique: A Manual for Its Planning and Implementation*. <http://www.tiu.edu/psychology/Twelker/critical_incident_technique.htm>.

Whittingham, R.B. *The Blame Machine: Why Human Error Causes Accidents*. Burlington, MA: Elsevier Butterworth-Heinemann, 2004.

6 PREVENTION THROUGH DESIGN
ADDRESSING OCCUPATIONAL RISKS IN THE DESIGN AND REDESIGN PROCESSES

ORIGINALLY PUBLISHED OCTOBER 2008

Transformative. That's what some have suggested could be the long-term impact of NIOSH's Prevention Through Design (PTD) initiative. Some believe it will lead to a fundamental shift in the practice of safety resulting in greater emphasis being given to the higher and more effective decision-making levels in the hierarchy of controls.

The goal of this initiative, founded on the need to "create a sustainable national strategy for prevention through design," is to "reduce the risk of occupational injury and illness by integrating decisions affecting safety and health in all stages of the design process." To move toward fulfillment of this mission, John Howard, M.D., 2002-08 director of NIOSH, said, "One important area of emphasis will be to examine ways to create a demand for graduates of business, architecture and engineering schools to have basic knowledge in occupational health and safety principles and concepts."

The PTD initiative is based on the premise that "one of the best ways to prevent and control occupational injuries, illnesses and fatalities is to design out or minimize hazards and risks early in the design process" (NIOSH). Notice that this definition limits activity to "early in the design process." At a July 2007 workshop that brought key PTD stakeholders together, many participants called for the concept to be extended to include redesign activities, much as the following definition does:

PTD: Addressing occupational safety and health needs in the design and redesign processes to prevent or minimize the work-related hazards and risks associated with the construction, manufacture, use, maintenance and disposal of facilities, materials, equipment and processes.

Enthusiasm for additional knowledge of PTD principles and practices was significant. Several workshop attendees said it would be helpful if a regulation or a standard were available that sets forth the principles and the methodologies to address hazards and risks in the design and redesign processes. The probability that OSHA could promulgate a regulation or a standard on PTD is unlikely at this time. It is more probable that an ANSI standard could be developed and approved, but that could take several years. For example, ANSI/AIHA Z10-2005 was published 6 years after the secretariat received ANSI approval to begin its work.

Let's assume that the NIOSH initiative, which is a several-year undertaking, is successful. Since hazards analyses and risk assessments are the core of the PTD concept, the impact on the knowledge needs of SH&E practitioners will be significant. As a primer, this article provides guidelines for addressing those needs.

At all levels—management, engineers, safety professionals—it must also be understood that safety standards and guidelines now include more provisions for addressing hazards and risks in the design and redesign processes. Examples of such standards and guidelines include the following:

•ANSI/ASSE Z241.1-2003, Control of Hazardous Energy: Lockout/Tagout and Alternative Methods (ASSE, 2003).

•ANSI/AIHA Z10-2005, Occupational Health and Safety Management Systems (ANSI/AIHA, 2005).

•ANSI/PMMI B155.1-2006, Safety Requirements for Packaging Machinery and Packaging-Related Converting Machinery (ANSI/PMMI, 2006).

•ANSI/RIA R15.06-1999, American National Standard for Industrial Robots and Robot Systems: Safety Requirements (Robotics Industries Association, 1999).

•*Aviation Ground Operation Safety Handbook* (6th ed.) (NSC, 2007).

•B11.TR3, Risk Assessment and Reduction: A Guideline to Estimate, Evaluate and Reduce Risks Associated With Machine Tools (ANSI/AMT, 2000).

•CSA Z1000-06, Occupational Health and Safety Management (Canadian Standards Association, 2006).

•ISO 14121, Safety of Machinery: Principles for Risk Assessment (ISO, 1999).

•ISO 12100-1, Safety of Machinery: Basic Concepts, General Principles for Design—Part 1 (ISO, 2003).

- SEMI S2-0706, Environmental, Health and Safety Guideline for Semiconductor Manufacturing Equipment (SEMI, 2006).
- SEMI S10-1103, Safety Guideline for Risk Assessment and Risk Evaluation Process (SEMI, 2003).

Promoting the acquisition of knowledge of safety through design/PTD concepts is also in concert with ASSE's position paper on designing for safety. The opening paragraph of that document states:

> Designing for safety (DFS) is a principle for design planning for new facilities, equipment and operations (public and private) to conserve human and natural resources and, thereby, protect people, property and the environment. DFS advocates systematic process to ensure state-of-the-art engineering and management principles are used and incorporated into the design of facilities and overall operations to ensure safety and health of workers, as well as protection of the environment and compliance with current codes and standards (ASSE, 1994).

Scope and Purpose

This article provides guidance on incorporating decisions pertaining to occupational risks into the design and redesign processes, including consideration of the life cycle of facilities, materials, equipment and processes. The goals of applying PTD principles are to:

- achieve safety, which is defined as that state for which the risks are acceptable and tolerable in the setting being considered;
- minimize the occurrence of occupational injuries, illnesses and fatalities.

Since this article is prompted by a NIOSH initiative and since NIOSH is exclusively an occupational safety and health entity, the scope of this article relates principally to the elimination, reduction or control of occupational risks. However, one cannot ignore the fact that the events or exposures which could result in occupational injuries and illnesses can also damage property and the environment, and interrupt business; those additional loss potentials are referred to in several places. In addition, the definition of safety through design—a broader definition than that of PTD—is included in the following list of definitions.

Safety Through Design Key Terms

Acceptable risk. That risk for which the probability of a hazards-related incident or exposure occurring and the severity of harm or damage that may result are as low as reasonably practicable and tolerable in the setting being considered. (This definition incorporates the ALARP concept.)

ALARP. That level of risk which can be further lowered only by an increment in resource expenditure that cannot be justified by the resulting decrement of risk.

Design. The process of converting an idea or market need into the detailed information from which a product or technical system can be produced.

Hazard. The potential for harm. Hazards include all aspects of technology and activity that produce risk. Hazards include the characteristics of things (e.g., equipment, dusts) and the actions or inactions of people.

Hazard analysis. A process that commences with recognition of a hazard and proceeds into an estimate of the severity of harm or damage that could result if its potential is realized and a hazard-related incident or exposure occurs.

Hierarchy of controls. A systematic way of thinking and acting, considering steps in a ranked and sequential order, to choose the most effective means of eliminating or reducing hazards and the risks that derive from them.

Life cycle. The phases of the facility, equipment, material and processes, including design and construction, operation, maintenance and disposal.

Prevention through design. Addressing occupational safety and health needs in the design and redesign processes to prevent or minimize the work-related hazards and risks associated with the construction, manufacture, use, maintenance, and disposal of facilities, materials, equipment and processes.

Probability. The likelihood of an incident or exposure occurring that could result in harm or damage—for a selected unit of time, events, population, items or activity being considered.

Residual risk. The risk remaining after preventive measures have been taken. No matter how effective the preventive actions, residual risk will always be present if a facility or operation continues to exist.

Risk. An estimate of the probability of a hazards-related incident or exposure occurring and the severity of harm or damage that could result.

Risk assessment. A process that commences with hazard identification and analysis, through which the probable severity of harm or damage is established, and concludes with an estimate of the probability of the incident or exposure occurring.

Safety. That state for which the risks are acceptable and tolerable in the setting being considered.

Safety through design. The integration of hazard analysis and risk assessment methods early in the design and redesign processes and taking the actions necessary so that the risks of injury or damage are at an acceptable level. This concept encompasses facilities, hardware, equipment, tools, materials, layout and configuration, energy controls, environmental concerns and products.

Severity. The extent of harm or damage that could result from a hazards-related incident or exposure.

Application

While these guidelines are applicable to all occupational settings, the focus is on providing assistance to managers, design engineers and safety professionals in smaller organizations (i.e., 1,000 or fewer employees). These guidelines apply to the three major timeframes in the practice of safety:

1) preoperational, in the design process, where the opportunities are greatest and the costs are lower for hazard and risk avoidance, elimination or control;

2) operational mode, where hazards are to be eliminated or controlled and risks reduced before their potentials are realized and hazards-related incidents or exposures occur;

3) postincident, as hazards-related incidents and exposures are investigated to determine causal factors and necessary risk-reduction measures.

Responsibility

Location management must provide the leadership to institute and maintain a policy and procedures affecting the design and redesign processes through which several goals are accomplished:

- Hazards are identified and analyzed.
- Risks deriving from the identified hazards are assessed and prioritized.
- Risks are reduced to an acceptable level through the application of the hierarchy of controls (see discussion starting on p. 116).

These methods are to be applied when new facilities, equipment and processes are acquired; when existing facilities, equipment and processes are altered; and when incidents are investigated.

All who have design responsibilities, as well as the operations personnel who will be affected and SH&E professionals should be involved in the decision-making process. In executing these responsibilities, management may:

- designate qualified in-house personnel to identify and analyze hazards, and assess the risks deriving from them for operations in place;
- employ independent consultants with hazard identification/analysis and risk assessment capabilities to assist with respect to operations in place and in the acquisition of new facilities, equipment, materials or processes;
- enter into arrangements with suppliers of newly acquired facilities, equipment, materials or processes to fulfill these responsibilities.

Relationships With Suppliers

Many organizations do not have design or technical staffs to fulfill the highlighted responsibilities. Thus, to avoid bringing hazards and risks into the workplace when new facilities, equipment, materials and processes are considered, and to ensure that hazards and risks are properly addressed when existing operations are altered, location management should take the following steps:

1) Establish design specifications and objectives.

2) Have an in-depth dialogue with suppliers and contractors on the expected use of the facilities, equipment and processes.

3) Include specifications to minimize bringing hazards and their related risks into the workplace in purchasing agreements and contracts for services.

4) Ask suppliers of services to attest that processes have been applied to identify and analyze hazards and to reduce the risks deriving from those hazards to an acceptable level. [There is precedent for having suppliers attest that risk analyses have been completed. Manufacturers of equipment to be used in the European Union are required by International Organization for Standardization (ISO) standards to certify that they have met applicable standards, including ISO 12100-1 and ISO 14121.]

5) Arrange for staff members (e.g., design engineers, SH&E professionals, maintenance personnel) to visit the supplier of equipment that the staff may consider hazardous, or for which design specifications have been provided to the supplier, before the equipment is shipped to ensure that safety needs have been met.

6) Require that a test run of the equipment is conducted during that visit.

7) Have an additional validation test performed after the equipment has been installed during which safety personnel or others sign off indicating that safety needs have been met.

Conducting Hazards Analyses and Risk Assessments

For many hazards and their associated risks, knowledge gained by management personnel, design engineers and safety professionals through education and experience will lead to proper conclusions on how to attain an acceptable risk level without bringing teams of people together for discussion. For more complex risk situations, however, it is vital to seek the counsel of experienced personnel at all levels who are close to the work or process. Reaching group consensus is a highly desirable goal. Sometimes, for what an SH&E professional considers obvious, achieving consensus is still desirable so that buy-in is obtained for the actions to be taken.

The goal of the risk assessment process, and the subsequent remediation actions, is to achieve acceptable risk levels. The risk assessment and remediation processes are not complete until acceptable risk levels are achieved. Other published standards or guidelines will be considered in the application of these guidelines.

However, applying existing standards may or may not attain acceptable risk levels. Standards offer only minimum requirements or may not contain provisions relating to the hazards in a given situation. Also, as they age, standards may become obsolete and inadequate in relation to more recently developed knowledge.

For example, designing lockout/tagout (LOTO) systems that meet all requirements of the National Electric Code, OSHA standards and ANSI/ASSE Z244.1-2003 may still result in unacceptable risk levels. In the analyses of electrocutions, one causal factor often found is that the LOTO station was inconveniently placed (e.g., 200 ft away, on the floor above), resulting in error-provocative and error-inviting situations. The standards cited do not require that LOTO stations be placed conveniently in the areas where the work is being performed.

A supplementary document to SEMI S2-0706 is titled "Related Information 1: Equipment/Product Safety Program." This document supports the premise that sometimes one must go beyond issued safety standards in the design process.

> Compliance with design-based safety standards does not necessarily ensure adequate safety in complex or state-of-the-art systems. It often is necessary to perform hazard analyses to identify hazards that are specific with the system, and develop hazard control measures that adequately control the associated risks beyond those that are covered in existing design-based standards (SEMI, 2006).

Although participants in the hazard analysis and risk assessment process will refer to existing standards as resources, the primary goal is to attain acceptable risk levels. A general guide on how to conduct a hazard analysis and how to extend the process into a risk assessment is offered in the following discussion. Whatever the simplicity or complexity of the hazard/risk situation, and whatever analysis method is used (there are many), the thought and action process outlined here is applicable.

Hazard Analysis and Risk Assessment Process

Although the focus is on eliminating, reducing and controlling occupational risks, as noted, this process is equally applicable in avoiding injury to the public, property and environmental damage, business interruption and product liability.

Establish Analysis Parameters

Select a manageable task, system, process or product to be analyzed, and establish its boundaries and operating phase (e.g., standard operation, maintenance, startup). Define its interface with other tasks or systems, if appropriate. Determine the scope of the analysis in terms of what can be harmed or damaged—people (employees, the public); property; equipment; productivity; the environment.

Identify the Hazards

A frame of thinking should be adopted that gets to the bases of causal factors, which are hazards. These questions would be asked: What aspects of technology or activity produce risk? What characteristics of things (equipment, dusts) or the actions or inactions of people present a potential for harm? Depending on the complexity of the hazardous situation, some or all of the following may apply:

- Use intuitive engineering and operational sense. This is paramount throughout.
- Examine system specifications and expectations.
- Review relevant codes, regulations and consensus standards.
- Interview current or intended system users or operators.
- Consult checklists.
- Review studies from similar systems.
- Consider the potential for unwanted energy releases.
- Account for possible exposures to hazardous environments.
- Review historical data (e.g., industry experience, incident investigation reports, OSHA and National Safety Council data, manufacturer's literature).
- Brainstorm.

Consider Failure Modes

Define the possible failure modes that would result in realization of the potentials of the hazards. Consider intentional and foreseeable misuse of facilities, equipment, materials and processes. Ask several questions: What circumstances can arise that would result in the occurrence of an undesirable event? What controls are in place that would mitigate the occurrence of such an event or exposure? How effective are the controls? Can controls be maintained easily? Can controls be defeated easily?

Determine Exposure Frequency and Duration

For each harm or damage category selected for the scope of the analysis (e.g., people, property, business interruption), estimate the frequency and duration of exposure to the hazard. This is an important part of this exercise. For instance, in a workplace situation, more judgments than one might realize will be made in this process. Ask, How often is a task performed? How long is the exposure period? How many people are exposed? What property or aspects of the environment are exposed?

Assess the Severity of Consequences

On a subjective basis, the goal is to identify the worst credible consequences should an incident occur, not the worst conceivable consequences. Historical data can be of great value as a baseline. Informed speculations are made to establish the consequences of an incident or exposure. Consider the following:

- number of injuries or illnesses and their severity, and fatalities that might occur;
- value of property or equipment that could be damaged;
- time for which the business may be interrupted and productivity lost;
- extent of environmental damage that could occur.

When the severity of the outcome of a hazards-related incident or exposure is determined, a hazard analysis has been completed.

Determine Occurrence Probability

Extending the hazard analysis into a risk assessment requires an additional step—estimating the likelihood (the probability) of a hazardous event or exposure occurring. Unless empirical data are available, which is rare, this is a subjective process. For more complex hazardous situations, it is necessary to brainstorm with knowledgeable people. To be meaningful, probability must be related to an interval base of some sort, such as a unit of time or activity, events, units produced, or the life cycle of a facility, equipment, process or product.

Define the Risk

It is necessary to conclude with a statement that contains the:
- probability of a hazards-related incident or exposure occurring;
- expected severity of adverse results;
- risk category (e.g., high, serious, moderate, low).

A risk assessment matrix should be used to identify risk categories. A matrix helps one communicate risk levels to decision makers (see Table 1 and Tables 2 and 3).

Rank Risks in Priority Order

A risk-ranking system should be adopted so that priorities can be established. Since the risk assessment exercise is subjective, this system would also be subjective. Prioritizing risks gives management the knowledge needed to appropriately allocate resources for their elimination or reduction.

Develop Remediation Proposals

When results of the risk assessment indicate that risk elimination, reduction or control measures are to be taken to achieve acceptable risk levels, several actions are needed:
- Alternate proposals for the design and operational changes necessary to achieve an acceptable risk level would be recommended.
- The actions shown in the hierarchy of controls (see discussion starting on p. 116) would be the base on which remedial proposals are made, in the order of their effectiveness.

- Remediation cost for each proposal would be determined and its effectiveness in achieving risk reduction would be estimated.
- Risk elimination or reduction methods would be selected and implemented to achieve an acceptable risk level.

Follow Up on Actions Taken

Good management requires that the effectiveness of actions taken to attain acceptable risk levels be assessed. Follow-up activity would establish that the:

- hazard/risk problem was resolved, only partially resolved or not resolved, as well as whether the actions taken created new hazards;
- risk should be reevaluated and other countermeasures proposed if the risk level achieved is not acceptable or if new hazards have been introduced.

Document the Results

Documentation, whether compiled under the direction of site management or by the equipment/service provider, should include comments on the:

- risk assessment method(s) used;
- hazards identified and the risks deriving from those hazards;
- reduction measures taken to attain acceptable risk levels.

Residual Risk

Residual risk is that which remains after preventive measures have been taken. No matter how effective the preventive actions, residual risk will always remain if an activity continues. Attaining zero risk is not possible. If the residual risk is not acceptable, the action outline presented in the hazard analysis and risk assessment process would be applied again. (Figure 1 provides an outline of one company's risk assessment process.)

Risk Assessment Matrixes

A risk assessment matrix provides a method to categorize combinations of probability of occurrence and severity of harm, thus establishing risk levels. A matrix helps one communicate about risk reduction actions with decision makers. Also, a matrix can be used to compare and prioritize risks, and to effectively allocate mitigation resources. It should be understood that definitions of terms used for incident probability and severity and for risk levels vary greatly in the many matrixes in use. Thus, an organization should create and obtain broad approval for a matrix that is suitable to the hazards and risks inherent in its operations.

Chapter 6: Prevention Through Design

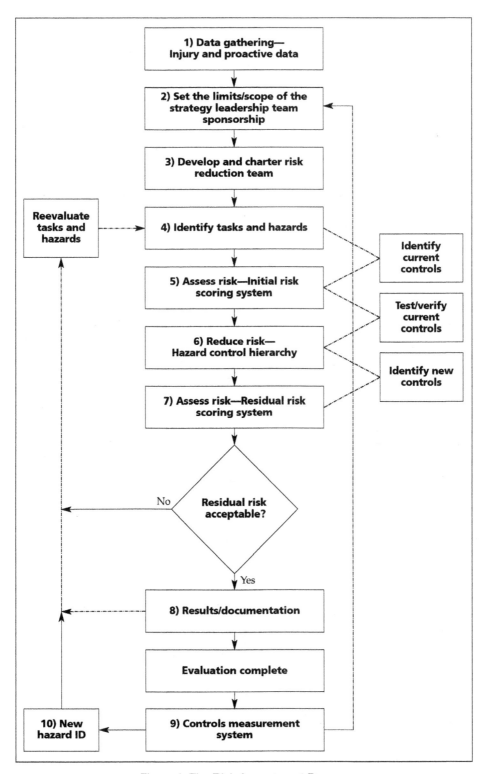

Figure 1: The Risk Assessment Process

Three examples of risk assessment matrixes are provided. Table 1 is adapted from a matrix in MIL-STD-882 D, Department of Defense Standard Practice for System Safety. (MIL-STD-882, first issued in 1969, is the grandfather of risk assessment matrixes.)

Table 2 is a composite of matrixes that include numerical values for probability and severity levels that are transposed into risk scorings. It is presented here for those who prefer to deal with numbers rather than qualitative indicators. (A word of caution for Table 2: The numbers are arrived at judgmentally and are qualitative.)

Table 3 is taken from ANSI B11.TR3-2000, the ANSI technical report titled "Risk Assessment and Risk Reduction—A Guide to Estimate, Evaluate and Reduce Risks Associated with Machine Tools." It is the base document used when the risk assessments shown in Figure 2 were conducted.

OCCURRENCE PROBABILITY	SEVERITY OF CONSEQUENCES			
	Catastrophic	Critical	Marginal	Negligible
Frequent	High	High	Serious	Medium
Probable	High	High	Serious	Medium
Occasional	High	Serious	Medium	Low
Remote	Serious	Medium	Medium	Low
Improbable	Medium	Medium	Medium	Low

Table 1: Risk Assessment Matrix

Severity levels and values	OCCURRENCE PROBABILITIES AND VALUES				
	Frequent (5)	Likely (4)	Occasional (3)	Seldom (2)	Unlikely (1)
Catastrophic (5)	25	20	15	10	5
Critical (4)	20	16	12	8	4
Marginal (3)	15	12	9	6	3
Negligible (2)	10	8	6	4	2
Insignificant (1)	5	4	3	2	1

Table 2: Risk Assessment Matrix: Numerical Gradings
Note: Numbers are arrived at judgmentally and are qualitative. > 15 = very high risk; 9 to 14 = high risk; 4 to 8 = moderate risk; < 4 = low risk

OCCURRENCE PROBABILITY	SEVERITY OF HARM			
	Catastrophic	Serious	Moderate	Minor
Very likely	High	High	High	Medium
Likely	High	High	Medium	Low
Unlikely	Medium	Medium	Low	Negligible
Remote	Low	Low	Negligible	Negligible

Table 3: Risk Assessment Matrix in ANSI B11.TR3-2000
Adapted from "Risk Assessment and Reduction: A Guide to Estimate, Evaluate and Reduce Risks Associated with Machine Tools (B11.TR3-2000)," by ANSI/AMT, 2000, McLean, VA: Authors.

Chapter 6: Prevention Through Design

designsafe Report Provided by design safety engineering, inc. www.designsafe.com

Application:	Transfer Line, Machine #334185
Description:	Sample assessment for demonstration
Analyst Name(s):	Steve, Rick, Rob plant operators, Bruce Jones, safety, Tom Woods, engineering
Company:	ABC Company
Facility Location:	Washington, DC
Product Identifier:	Model 89RX-1
Assessment Type:	Detailed
Limits:	This initial risk assessment is for certain Operator tasks
Sources:	on site investigations, discussions w/ plant personnel
Risk Scoring System:	ANSI B11 TR3 Two Factor

Guide sentence: When doing [task], the [user] could be injured by the [hazard] due to the [failure mode].

			Initial Assessment			Final Assessment		
Item Id	User / Task	Hazard / Failure Mode	Severity Probability	Risk Level	Risk Reduction /Comments	Severity Probability	Risk Leve	Status / Responsible / /Reference
1-1-1	operator(s) tool change	mechanical : cutting / severing	Moderate Remote	Negligible	gloves /issue to all new hires	Minor Remote	Negligible	Complete Joe
1-1-2	operator(s) tool change	mechanical : impact dropping heavy tool	Moderate Unlikely	Low	lift assist, standard procedures	Minor Remote	Negligible	Complete
1-1-3	operator(s) tool change	mechanical : pinch points	Minor Remote	Negligible	standard procedures	Minor Remote	Negligible	Complete
1-1-4	operator(s) tool change	mechanical : head bump on overhead objects	Minor Remote	Negligible	other	Minor Remote	Negligible	Complete
1-1-5	operator(s) tool change	ergonomics / human factors : lifting / bending / twisting	Minor Remote	Negligible	look into lift assists or quick release fasteners	Minor Remote	Negligible	Complete Jane
1-1-6	operator(s) tool change	slips / trips / falls : slips	Serious Likely	High	graded floors, non-slip flooring, contain coolant, footwear	Minor Remote	Negligible	In-process Jane
1-2-1	operator(s) remove reject parts	mechanical : cutting / severing	Moderate Remote	Negligible		Minor Remote	Negligible	Complete
1-2-2	operator(s) remove reject parts	mechanical : drawing-in / trapping	Catastrophic Likely	High	interlocked barriers, presence sensing devices, stop line to pull part /requisition submitted	Minor Remote	Negligible	In-process John
1-2-3	operator(s) remove reject parts	mechanical : impact by dropped parts	Moderate Unlikely	Low		Minor Remote	Negligible	Complete
1-2-4	operator(s) remove reject parts	ergonomics / human factors : lifting / bending / twisting	Moderate Remote	Negligible		Minor Remote	Negligible	Complete
1-3-1	operator(s) probe check	Other : None, no hazards						Complete

Figure 2: Sample Initial Hazard Analysis and Risk Assessment Worksheet

Management Decision Levels

Remedial action or acceptance levels must be attached to the risk categories to permit intelligent management decision making. Table 4 provides a basis for review and discussion. Others who craft risk assessment matrixes may have differing ideas about acceptable risk levels and the management actions to be taken in a given risk situation.

Risk category	Remedial action or acceptance
High	Operation not permissible
Serious	Remedial action to have high priority
Medium	Remedial action to be taken in appropriate time
Low	Risk is acceptable: remedial action discretionary

Table 4: Management Decision Levels

Selecting Probability and Severity

There is no one right method for selecting probability and severity categories and their descriptions, and many variations are in use. Examples in Tables 5 through 8 show variations in the terms and their descriptions as used in applied risk assessment processes for probability of occurrence and severity of consequence.

Hazards Analysis and Risk Assessment Techniques

Over the past 40 years, a large number of hazard analysis and risk assessment techniques have been developed. Clemens (1982) gives brief descriptions of 25 techniques, while Stephans and Talso (1997) describe 101 methods. Brief descriptions of select hazard analysis techniques are offered here. As a practical matter, having knowledge of three techniques—initial hazard analysis and risk assessment, the what-if/checklist analysis methods, and failure modes and effects analysis—will be sufficient to address most risk situations.

It is important to understand that each of these techniques complements, rather than supplants, the others. Selecting the technique or a combination of techniques used to analyze a hazardous situation requires good judgment based on knowledge and experience. Qualitative rather than quantitative judgments will prevail. For all but the complex risks, qualitative judgments will be sufficient. Sound quantitative data on incident and exposure probabilities are seldom available. Many quantitative risk assessments are really qualitative risk assessments because so many judgments have to be made when deciding on the probability levels.

Descriptive word	Probability description
Frequent	Likely to occur repeatedly
Probable	Likely to occur several times
Occasional	Likely to occur sometime
Remote	Not likely to occur
Improbable	So unlikely can assume occurrence will not be experienced

Table 5: Probability Descriptions: Example A

Descriptive word	Probability description
Frequent	Could occur annually
Likely	Could occur once in 2 years
Possible	Not more than once in 5 years
Rare	Not more than once in 10 years
Unlikely	Not more than once in 20 years

Table 6: Probability Descriptions: Example B

Descriptive word	Severity description
Catastrophic	Death or permanent total disability, system loss, major property damage and business downtime
Critical	Permanent, partial, or temporary disability in excess of 3 months, major system damage, significant property damage and downtime
Marginal	Minor injury, lost workday accident, minor system damage, minor property damage and little downtime
Negligible	First aid or minor medical treatment, minor system impairment

Table 7: Severity Descriptions for Multiple Harm & Damage Categories: Example A

Descriptive word	Severity description
Catastrophic	One or more fatalities, total system loss, chemical release with lasting environmental or public health impact
Critical	Disabling injury or illness, major property damage and business down time, chemical release with temporary environmental or public health impact
Marginal	Medical treatment or restricted work, minor subsystem loss or damage, chemical release triggering external reporting requirements
Negligible	First aid only, nonserious equipment or facility damage, chemical release requiring only routine cleanup without reporting

Table 8: Severity Descriptions for Multiple Harm & Damage Categories: Example B

Preliminary Hazard Analysis:
Initial Hazard Analysis and Risk Assessment

The preliminary hazard analysis technique has its origins in system safety. It is used to identify and evaluate hazards in the very early stages of the design process. In actual practice, however, the technique has attained much broader use. The principles on which a preliminary hazard analysis are based are used not only in the initial design process, but also in assessing the risks of existing operations or products. Thus, the technique needs a new name: initial hazard analysis and risk assessment.

Headings on initial hazard analysis forms will include typical identification data such as date, evaluators' names, the department and location. The following information is usually included in an initial hazard analysis process:

- hazard description (also called a hazard scenario);
- description of the task, operation, system, subsystem or product being analyzed;
- exposures to be analyzed: people (employees, the public); facility, product or equipment loss; operation downtime; environmental damage;
- probability interval to be considered: unit of time or activity; events; units produced; life cycle;
- risk assessment code, using the agreed upon risk assessment matrix;
- remedial action to be taken, if risk reduction is needed.

A communication accompanies the analysis, indicating the assumptions made and the rationale for them. Comment would be made on the assignment of responsibilities for the remedial actions to be taken and expected completion dates. (Figure 2 presents a sample of an initial hazard analysis and risk assessment worksheet.)

What-If Analysis

For a what-if analysis, a group of people (as few as two, but often several more) use a brainstorming approach to identify hazards, hazard scenarios, failure modes, how incidents can occur and their probable consequences. Questions posed during this session may commence with what-if, as in "What if the air conditioning fails in the computer room?" or may express general concerns, as in "I worry about the possibility of spillage and chemical contamination during truck off-loading."

All questions are recorded and assigned for investigation. Each subject of concern is then addressed by one or more team members. They would consider the potential of the hazardous situation and the adequacy of risk controls in effect, suggesting additional risk reduction measures if appropriate.

Chapter 6: Prevention Through Design

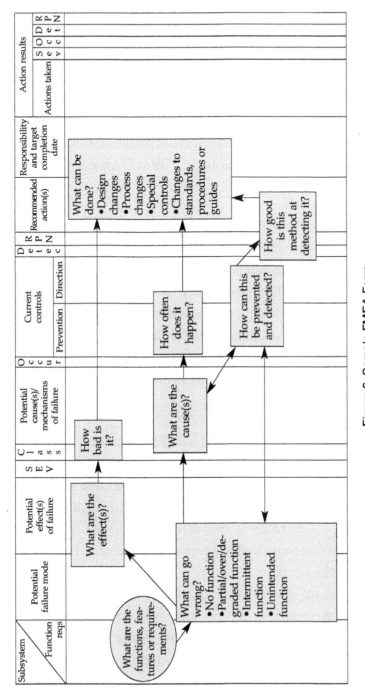

Figure 3: Sample FMEA Form
(Note. Adapted from *Potential Failure Mode and Effects Analysis* (3rd ed.), by Automotive Industry Action Group (AIAG), 2001, Southfield, MI: Author.)

Checklist Analysis

Checklists are primarily adaptations from published standards, codes and industry practices. There are many such checklists. They consist of questions pertaining to the applicable standards and practices—usually with a yes, no or not applicable response. Their purpose is to identify deviations from the expected and, thereby, possible hazards. A checklist analysis requires a walkthrough of the area to be surveyed. Checklists are easy to use and provide a cost-effective way to identify customarily recognized hazards.

However, the quality of the checklists depends on the experience of those who develop them. Furthermore, they must be crafted to suit particular facility/operations needs. If a checklist is not complete, the analysis may not identify some hazardous situations.

What-If/Checklist Analysis

The what-if/checklist hazard analysis technique combines the creative, brainstorming aspects of the what-if method with the systematic approach of a checklist. Combining the techniques can compensate for the weaknesses of each. The what-if part of the process can help the team identify hazards that have the potential to be causal factors for incidents, even though no such incidents have yet occurred. The checklist segment provides a systematic review that can serve as an idea generator during the what-if brainstorming process. Usually, a team experienced in the operation's design, operation and maintenance performs the analysis.

Hazard and Operability Analysis

The hazard and operability analysis (HAZOP) technique was developed to identify both hazards and operability problems in chemical process plants. It has subsequently been applied to a wide range of industry processes and equipment. An interdisciplinary team and an experienced team leader are required. In a HAZOP application, a process or operation is systematically reviewed to identify deviations from desired practices that could lead to adverse consequences. HAZOPs can be used at any stage in the life of a process.

A HAZOP usually requires prework in gathering materials and a series of meetings in which the team, using process drawings, systematically evaluates the impact of deviations from the desired practices. The team leader uses a set of guidewords to develop discussions. As the team reviews each step in a process, several items are documented including:

•deviations and their causal factors;
•consequences should an incident occur;

- safeguards in place;
- required actions or the need for more information to evaluate the deviation.

Failure Modes and Effects Analysis

In several industries, failure modes and effects analyses (FMEAs) have been the techniques of choice by design engineers for reliability and safety considerations. They are used to evaluate the ways in which equipment fails and the response of the system to those failures. Although an FMEA typically occurs early in the design process, the technique can also serve well as an analysis tool throughout the life of equipment or a process.

An FMEA produces qualitative, systematic lists that include the failure modes, the effects of each failure, safeguards that exist and additional actions that may be necessary. For example, for a pump, the failure modes would include failure to stop when required; stops when required to run; seal leaks or ruptures; and pump case leaks or ruptures.

Both the immediate effects and the impact on other equipment would be documented. Generally, when analyzing impacts the probable worst-case scenario is assumed and analysts would determine whether existing safeguards are adequate. Although an FMEA can be performed by one person, a team is typically appointed when there is complexity. In either case, the process follows a similar path:
- Identify the item or function to be analyzed.
- Define the failure modes.
- Document the failure causes.
- Determine the failure effects.
- Assign a severity code and a probability code for each effect.
- Assign a risk code.
- Record the actions required to reduce the risk to an acceptable level.

The FMEA process requires entry of probability, severity and risk codes. Figure 3 presents a sample FMEA form on which those codes would be entered. Good references explaining risk coding for FMEA purposes include *Potential Failure Mode and Effects Analysis* (AIAG, 2001) and *Failure Mode and Effects Analysis: A Guide for Continuous Improvement* (International SEMATECH, 1992).

Fault Tree Analysis

A fault tree analysis (FTA) is a top-down, deductive logic model that traces the failure pathways for a predetermined, undesirable condition or event, called the top event. An FTA can be conducted either quantitatively or subjectively. A subjective (or qualitative) analysis can produce suitable results, especially when quantitative numbers are not available. The FTA generates a fault tree (a symbolic logic model) entering failure probabilities for the combinations of equipment failures and human errors that can result in the accident. Each immediate causal factor is examined to determine its subordinate causal factors until the root causal factors are identified.

The strength of an FTA is its ability to identify combinations of basic equipment and human failures that can lead to an incident, allowing the analyst to focus preventive measures on significant basic causes. An FTA has particular value when analyzing highly redundant systems and high-energy systems in which high-severity events can occur.

For systems vulnerable to single failures that can lead to accidents, the FMEA and HAZOP techniques are better suited. FTA is often used when another technique has identified a hazardous situation that requires a more detailed analysis. Conducting an FTA of other than the simplest systems requires the talent of experienced analysts.

Management Oversight and Risk Tree

All of the hazard analysis and risk assessment techniques previously discussed relate principally to the initial design process in the preoperational mode, and to the redesign process to achieve risk reduction in the operational mode. The management oversight and risk tree (MORT) is relative to the postincident time frame in the practice of safety. MORT was developed principally for incident investigations. U.S. Department of Energy (1994) describes MORT as follows:

> MORT is a comprehensive analytical procedure that provides a disciplined method for determining the systemic causes and contributing factors of accidents. MORT directs the user to the *hazards and risks deriving from both system design and procedural shortcomings* (emphasis added).

MORT provides an excellent resource for postincident investigations. Investigation results may prompt use of the hazard identification and analysis and risk assessment methods described.

Hierarchy of Controls

A hierarchy of controls provides a systematic way of thinking, considering steps in a ranked and sequential order, to choose the most effective means of eliminating or

Chapter 6: Prevention Through Design

reducing hazards and their associated risks. Acknowledging that premise—that risk-reduction measures should be considered and taken in a prescribed order—represents an important step in the evolution of the practice of safety.

Achieving Acceptable Risk

In applying a hierarchy of controls, the desired outcome of actions taken is to achieve an acceptable risk level. Acceptable risk, as previously defined, is that risk for which the probability of a hazards-related incident or exposure occurring and the severity of harm or damage that could result are as low as reasonably practicable and tolerable in the situation being considered. That definition requires several factors be taken into consideration:

•avoiding, eliminating or reducing the probability of a hazards-related incident or exposure occurring;

•reducing the severity of harm or damage that may result if an incident or exposure occurs;

•the feasibility and effectiveness of risk-reduction measures to be taken, and their costs, in relation to the amount of risk reduction to be achieved.

Six Levels of Action

Decision makers should understand that with respect to the six levels of action shown in the following hierarchy of controls the methods described in the first, second and third action levels are more effective because they:

•are preventive actions that eliminate/reduce risk by design, substitution and engineering measures;

•rely the least on the performance of personnel;

•are less defeatable by supervisors or workers.

Actions described in the fourth, fifth and sixth levels are contingent actions and rely greatly on the performance of personnel for their effectiveness.

The following hierarchy of controls is considered state-of-the-art, and it is compatible with the hierarchy in ANSI/AIHA Z10-2005:

1) Eliminate or reduce risks in the design and redesign processes.
2) Reduce risks by substituting less hazardous methods or materials.
3) Incorporate safety devices.
4) Provide warning systems
5) Apply administrative controls (e.g., work methods, training, work scheduling).
6) Provide PPE.

The Logic of Taking Action in the Order Given

The following discussion addresses each action element in the hierarchy of controls, including providing a rationale for listing actions to be taken in the order given. Taking actions in the prescribed order, as feasible and practicable, is the most effective means to achieve risk reduction.

Eliminate or Reduce Risk in the Design and Redesign Processes

The theory is plainly stated. If hazards are eliminated in the design and redesign processes, risks that derive from those hazards are also eliminated. But, elimination of hazards completely by modifying the design may not always be practicable. In such cases, the goal is to modify the design, within practicable limits, so that the 1) probability of personnel making human errors because of design inadequacies is at a minimum; and 2) ability of personnel to defeat the work system and the work methods prescribed, as designed, is at a minimum. Examples would include designing to eliminate or reduce the risk from hazards related to falls, ergonomics, confined space entry, electricity, noise and chemicals.

Substitute Less-Hazardous Method/Material

Substitution of a less-hazardous method or material may also reduce the risks. However, substitution may or may not result in equivalent risk reduction as might occur if the hazards and risks were addressed through system design or redesign.

Consider this example. A mixing process for chemicals involves considerable manual materials handling. A reaction occurs and an employee sustains serious chemical burns. Identical operations are performed at two of the company's locations. At one, the operation is redesigned so that it is completely enclosed, automatically fed and operated by computer from a control panel, thus greatly eliminating operator exposure.

At the other location, redesign funds are not available. To reduce the risk, the supplier agrees to premix the chemicals before shipment (substitution). Some mechanical feed equipment for the chemicals is also installed. The risk reduction achieved by substitution is not equivalent to that attained by redesigning the operation, so additional administrative controls are required.

Methods that illustrate substituting a less-hazardous method, material or process include using automated materials handling equipment; providing an automatic feed system to reduce machine hazards; using a less-hazardous cleaning material; reducing speed, force or amperage; reducing pressure or temperature; replacing a dated steam heating system and its boiler explosion hazards with a hot-air system.

Incorporate Safety Devices

When safety devices are incorporated into the system in the form of engineering controls, substantial risk reduction can be achieved. Engineered safety devices are intended to prevent access to the hazard by workers—to separate hazardous energy from the worker and deter worker error. Examples include machine guards, interlock systems, circuit breakers, start-up alarms, presence-sensing devices, safety nets, ventilation systems, sound enclosures, fall prevention systems, and lift tables, conveyors and balancers.

Install Warning Systems

Warning system effectiveness relies considerably on administrative controls, such as training, drills, the quality of maintenance and the reactions of people. Although vital in many situations, warning systems may be reactionary in that they alert people only after a hazard's potential is in the process of being realized (e.g., a smoke alarm). Examples of warning systems include smoke detectors, alarm systems, backup alarms, chemical detection systems, signs and alerts in operating procedures or manuals.

Institute Administrative Controls

Administrative controls rely on the methods chosen being appropriate in relation to needs, the capabilities of those responsible for their delivery and application, the quality of supervision and the expected performance of the workers. Administrative controls include personnel selection, developing and applying appropriate work methods and procedures, training, supervision, motivation, behavior modification, work scheduling, job rotation, scheduled rest periods, maintenance, management of change, investigations and inspections.

Achieving a superior level of effectiveness in all of these administrative methods is difficult and not often accomplished.

Provide Personal Protective Equipment

The proper use of PPE relies on an extensive series of supervisory and personnel actions, such as identifying and selecting the type of equipment needed, proper fitting and training, inspecting and maintaining. Although PPE is necessary in many occupational situations, it is the least effective way to deal with hazards and risks because systems put in place for their use can be easily defeated. One goal of the design processes should be to reduce reliance on PPE to a practical minimum, applying the ALARP concept. PPE examples include safety glasses, face shields, respirators, welding screens, safety shoes, gloves and hearing protection.

The Descending Order of Controls

For many risk situations, a combination of the risk management methods in the hierarchy of controls is necessary to achieve acceptable risk levels. However, the expectation is that each step will be considered in descending order and that reasonable attempts will be made to eliminate or reduce hazards and their associated risks through steps higher in the hierarchy before lower steps are considered. A lower step in the hierarchy of controls is not to be chosen until practical applications of the preceding level or levels are exhausted.

A yet-to-be published document, MIL-STD-882E, includes provisions that further explain the governing thought processes when the hierarchy of controls is applied. Excerpts from this standard follow.

> System safety mitigation order of precedence as in 882E: In reducing risk, the cost, feasibility, and effectiveness of candidate mitigation methods should be considered. In evaluating mitigation effectiveness, an order of precedence generally applies as follows.
>
> a) Eliminate hazard through design selection: Ideally, the risk of a hazard should be eliminated. This is often done by selecting a design alternative that removes the hazard altogether.
>
> b) Reduce mishap risk through design alteration: If the risk of a hazard cannot be eliminated by adopting an alternative design, design changes should be considered that reduce the severity and/or the probability of a harmful outcome.
>
> c) Incorporate engineered safety features (ESF): If unable to eliminate or adequately mitigate the risk of a hazard through a design alteration, reduce the risk using an ESF that actively interrupts the mishap sequence.
>
> d) Incorporate safety devices: If unable to eliminate or adequately mitigate the hazard through design or ESFs, reduce mishap risk by using protective safety features or devices.
>
> e) Provide warning devices: If design selection, ESFs, or safety devices do not adequately mitigate the risk of a hazard, include a detection and warning system to alert personnel to the presence of a hazardous condition or occurrence of a hazardous event.
>
> f) Develop procedures and training: Where other risk reduction methods cannot adequately mitigate the risk from a hazard, incorporate special procedures and training. Procedures may prescribe the use of personal protective equipment (U.S. Department of Defense, 2005).

Conclusion

SH&E professionals cannot ignore the favorable impact of the PTD movement. It is becoming evident that the primary focus is risk—identifying and analyzing hazards, and assessing the risks deriving from them. The entirety of purpose of those responsible for safety is to manage their endeavors with respect to hazards so that their associated risks are acceptable. ■

References

ANSI/AIHA. (2005). Occupational health and safety management systems (ANSI/Z10-2005). Fairfax, VA: Authors.

ANSI/ASSE. (2003). Control of hazardous energy: Lockout/tagout and alternative methods (ANSI/ASSE Z241.1-2003). Des Plaines, IL: Authors.

ANSI/Association for Manufacturing Technology (AMT). (2000). Risk assessment and reduction: A guide to estimate, evaluate and reduce risks associated with machine tools (B11.TR3-2000). McLean, VA: Authors.

ANSI/Packaging Machinery Manufacturers Institute (PMMI). (2006). Safety requirements for packaging machinery and packaging-related converting machinery (B155.1-2006). Arlington, VA: Authors.

ANSI/Robotics Industries Association (RIA). (1999). American national standard for industrial robots and robot systems: Safety requirements (ANSI/RIA R15.06-1999). Ann Arbor, MI: Authors.

ASSE. (1994). Position paper on designing for safety. Des Plaines, IL: Author.

Automotive Industry Action Group (AIAG). (2001). *Potential failure mode and effects analysis: FMEA* (3rd ed.). Southfield, MI: Author.

Canadian Standards Association (CSA). (2006). Occupational health and safety management (CSA Z1000-06). Ottawa, Ontario: Author.

Center for Chemical Process Safety. (1992). *Guidelines for hazard evaluation procedures* (2nd ed. with worked examples). Hoboken, NJ: John Wiley & Sons.

Christensen, W.C. (2003, March). Safety through design: Helping design engineers answer 10 key questions. *Professional Safety, 48*(3), 32-39.

Christensen, W.C. (2007, May). Retrofitting for safety: Career implications for SH&E personnel. *Professional Safety, 52*(5), 36-44.

Christensen, W.C. & Manuele, F.A. (2000). *Safety through design*. Itasca, IL: National Safety Council.

Clemens, P. (1982, March/April). A compendium of hazard identification and evaluation techniques for system safety application. *Hazard Prevention, 2*(2), 11-18.

Flamberg, S., Leverenz, F., Rose, S., et al. (2007). Guidance document for incorporating risk concepts into NFPA codes and standards. Quincy, MA: The Fire Protection Research Foundation. Retrieved Sept. 2, 2008, from http://www.nfpa.org/assets/files//PDF/Research/Risk-base_Codes_and_Stds-Appendices.pdf.

International Organization for Standardization (ISO). (1999). Safety of machinery: Principles for risk assessment (ISO 14121). Geneva, Switzerland: Author.

ISO. (2003). Safety of machinery: Basic concepts, general principles for design—Part 1: Basic terminology, methodology (ISO 12100-1). Geneva, Switzerland: Author.

International SEMATECH. (1992). *Failure mode and effects analysis (FMEA): A guide for continuous improvement for the semiconductor equipment industry* (Technology Transfer No. 92020963A-ENG). Austin, TX: Author. Retrieved Sept. 2, 2008, from http://www.fmeainfocentre.com/handbooks/sematechsemiconductorfmeahandbook.pdf.

Main, B. (2002, July). Risk assessment is coming. Are you ready? *Professional Safety, 47*(7), 32-37.

Main, B. (2004). *Risk assessment: Basics and benchmarks.* Ann Arbor, MI: design safety engineering inc.

Main, B. (2004, Dec.). Risk assessment: A review of the fundamental principles. *Professional Safety, 49*(12), 37-47.

Manuele, F.A. (2003). *On the practice of safety* (3rd ed.). Hoboken, NJ: John Wiley & Sons.

Manuele, F.A. (2005, May). Risk assessments and hierarchies of control. *Professional Safety, 50*(5), 33-39.

Manuele, F.A. (2006, Feb.). ANSI/AIHA Z10-2005: The new benchmark for safety management systems. *Professional Safety, 51*(2), 25-33.

Manuele, F.A. (2007). *Advanced safety management: Focusing on Z10 and serious injury prevention.* Hoboken, NJ: John Wiley & Sons.

National Safety Council. (2007). *Aviation ground operation safety handbook* (6th ed.). Itasca, IL: Author.

NIOSH. Prevention through design (NIOSH Safety and Health Topic). Washington, DC: U.S. Department of Health and Human Services, Centers for Disease Control and Prevention, Author. Retrieved Sept. 3, 2008, from http://www.cdc.gov/niosh/topics/ptd.

Semiconductor Equipment and Materials International (SEMI). (2006). Environmental, health and safety guideline for semiconductor manufacturing equipment (SEMI S2-0706). San Jose, CA: Author.

Society of Fire Protection Engineers (SFPE). (2006). *SFPE engineering guide to fire risk assessment.* Bethesda, MD: Author.

Stephans, R. (2004). *System safety for the 21st century.* Hoboken, NJ: John Wiley & Sons.

Stephans, R. & Talso, W.W. (Eds.). (1997). *System safety analysis handbook: A sourcebook for safety practitioners.* Albuquerque, NM: System Safety Society, New Mexico Chapter.

Swain, A.D. (1963). *Work situation approach to improving job safety* (Report SC-R-69-1320). Albuquerque, NM: Sandia Laboratories.

U.S. Department of Defense. (2000). Standard practice for system safety (MIL-STD-882D). Washington, DC: Author. Retrieved Sept. 2, 2008, from http://www.safetycenter.navy.mil/instructions/osh/milstd882d.pdf#search='MILSTD882D.

U.S. Department of Defense. (2005). Draft standard practice for system safety (MIL-STD-882E). Washington, DC: Author. Retrieved Sept. 3, 2008, from http://www.system-safety.org/~casacramento/Standards.htm.

U.S. Department of Energy. (1994). *Guide to use of the management oversight and risk tree* (SSDC-103). Washington, DC: Author.

Vincoli, J.W. (1993). *Basic guide to system safety.* Hoboken, NJ: John Wiley & Sons.

7 SERIOUS INJURIES AND FATALITIES
A CALL FOR A NEW FOCUS ON THEIR PREVENTION

ORIGINALLY PUBLISHED DECEMBER 2008

Over the past few decades, serious injuries and workplace fatalities have been significantly reduced. However, statistical trends in the more recent past indicate that additional research and knowledge are needed about causation and preventive measures so that safety professionals can give counsel on how these injuries and fatalities can be further reduced. To achieve this, SH&E professionals must adopt a new mindset that gives serious injury prevention a higher priority.

This will require several actions. The safety professional must address the phenomenon that seems to have developed in companies which continue to report serious injuries and fatalities despite otherwise stellar performance. In addition, the myth that preventing incidents that occur frequently will equivalently encompass severity reduction must be debunked. Other factors, such as organizational safety culture with respect to preventing serious injuries and the effect of the current economic climate, must be considered as well.

To help SH&E professionals in these endeavors, a mechanism for an internal study of severity potential is provided and the need for improved incident investigation is emphasized. In addition, an outline is presented for conducting a gap analysis that would compare existing safety management systems to the provisions of ANSI/AIHA Z10.

The Safety Performance Phenomenon

In early 2007, the Alcoa Foundation awarded a grant to Indiana University of Pennsylvania (IUP) to support a national forum on fatality prevention in the workplace. In a news release (Alcoa, 2007) announcing the grant, Lon Ferguson, chair of IUP's Safety Sciences Department, said:

> The reliance on traditional approaches to fatality prevention has not always proven effective. This fact has been demonstrated by many companies, including some thought to be top performers in worker safety and health, as they continue to experience fatalities while at the same time achieving benchmark performance in reducing less serious injuries and occupational illnesses.

The author's analyses, made over the past several years, support Ferguson's statement. Traditional safety management systems may not adequately address severe injury and fatality potentials. Others have also recognized the phenomenon. For example, the membership of ORC Worldwide (formerly Organization Resource Counselors) consists of about 140 Fortune 500 companies. Many of those companies have outstanding safety cultures and commendable safety management systems in place.

However, because some of those companies continue to experience fatalities and serious injuries, ORC is creating a special system to gather data on the specifics of their occurrence. It is expected that the system will include fatalities, serious injuries that had fatality potential and near-hits that under other circumstances may have resulted in serious consequences. The data will be analyzed with the hope that the outcomes will provide more information than is now available for their prevention.

Collectively, SH&E professionals should ask whether there is adequate in-depth knowledge of the causal factors for low-probability/serious-consequence incidents. The author's research on incident investigations (see p. 129) suggests that there is not.

Statistical Indicators: Fatalities

The reduction in both the number of serious injuries and fatalities and their rates in recent years must be recognized, as they are an indication that those involved in the practice of safety are doing many things right. The fatality rate data in Tables 1 and 2 are based on excerpts from National Safety Council's *Accident Facts* (now *Injury Facts*) and the Bureau of Labor Statistics' (BLS) annual census of fatal occupational injuries. The fatality rate is the number of fatalities per 100,000 workers.

Years ending in 1 were chosen as a focal point for this review so that an observation could be made of results since OSHA took effect in 1971. While employment increased more than 280% from 1941 to 2001, the number of fatalities dropped more than 67%—and the fatality rate dropped more than 88%. This record is highly favorable and complimentary to all involved.

One also cannot ignore the emergence of OSHA in 1971 and the greater concentration on workplace safety that followed. Using 1971 data as a base, the fatality rate was reduced about 75% by 2001. Table 2 picks up from Table 1 and provides data on fatalities and fatality rates since 2001.

Year	No. of fatalities	Fatality rate	No. of workers (1,000s)
1941	18,000	37	48,100
1951	16,000	28	57,450
1961	13,500	21	64,500
1971	13,700	17	78,500
1981	12,500	13	99,800
1991	9,800	8	116,400
2001	5,900	4.3	136,000

Table 1: All Fatalities, All Occupations: 1941-2001
Note. Data based on excerpts from *Accident Facts*, by National Safety Council, 1995, Itasca, IL: Author, and "Census of Fatal Occupational Injuries, 1997-2006," by Bureau of Labor Statistics, 2007, Washington, DC: U.S. Department of Labor, Author. The fatality rate is the number of fatalities per 100,000 workers. Years ending in 1 were chosen as a focal point for this review so that an observation could be made of results since OSHA took effect in 1971.

Year	No. of fatalities	Fatality rate
2001	5,900	4.3
2002	5,524	4.0
2003	5,559	4.0
2004	5,703	4.1
2005	5,702	4.0
2006	5,703	3.9

Table 2: All Fatalities, All Occupations: 2001-2006
Note. Data based on excerpts from "Census of Fatal Occupational Injuries, 1997-2006," by Bureau of Labor Statistics, 2007, Washington, DC: U.S. Department of Labor, Author. The fatality rate is the number of fatalities per 100,000 workers.

According to the 2001 data, 5,900 fatalities occurred and the fatality rate was 4.3. However, consider the data for 2002 through 2006. The number of fatalities increased 3.2% and the fatality rate remained the same. Fatality rates over a 6-year period have not varied substantially.

Why did the number of fatalities increase? Has there been a reversal of the downward trend experienced in previous years? Why did the fatality rate not continue the remarkable reductions seen in the years from 1941 through 2001? Safety professionals have a responsibility to promote the causal factor research needed to answer those questions.

Statistical Indicators: Serious Injury Trending

Data on the characteristics of serious injuries and workers' compensation claims frequency have been extracted from two sources: National Council on Compensation Insurance (NCCI) and BLS.

National Council on Compensation Insurance

In 2006, NCCI issued a video, *The Remarkable Story of Declining Frequency—Down 30% in the Past Decade*. This 12-minute video reports that workers' compensation claim frequency is down considerably, not only in the U.S. but also in several industrialized countries.

However, a 2005 NCCI research brief titled "Workers' Compensation Claim Frequency Down Again," states that "there has been a larger decline in the frequency of smaller lost-time claims than in the frequency of larger lost-time claims."

Consider the trend numbers presented in Table 3, taken from NCCI's *State of the Line* report (Mealy, 2005). These data show reductions in selected categories of claim values for the years 1999 and 2003, expressed in 2003 hard dollars. While the frequency of workers' compensation cases is down, the greatest reductions are in lower cost claims. The reduction in cases valued from $10,000 to $50,000 is about one-third of that for cases valued at less than $2,000. For cases valued over $50,000, the reduction is about one-fifth of that for the less costly injuries. Thus, costly claims—those for serious injuries and fatalities—loom larger within the spectrum of all claims reported.

Value of claim	Declines in frequency
Less than $2,000	34%
$2,000 to $10,000	21%
$10,000 to $50,000	11%
More than $50,000	7%

Table 3: Categories of Injury Reductions

Bureau of Labor Statistics

For many years, BLS has issued reports titled *Lost Work-Time Injuries and Illnesses: Characteristics and Resulting Time Away From Work* and *Workplace Injuries and Illnesses*. The data in Table 4, taken from those reports, indicate that the total number of cases

Year	Total cases	DART rates
2000	1,664,018	3.0
2001	1,537,567	2.9
2002	1,436,200	2.8
2003	1,315,920	2.6
2004	1,259,320	2.5
2005	1,234,680	2.4
2006	1,183,500	2.3

Table 4: Trends for Lost-Worktime Cases
Note. Data from "Lost Worktime Injuries and Illnesses: Characteristics and Resulting Time Away From Work, 1995-2006," by Bureau of Labor Statistics, 2007. Washington, DC: U.S. Department of Labor, Author. DART rate includes cases with days away from work, job transfer or restriction.

	Percent of days away from work cases by number of days						
	1	2	3-5	6-10	11-20	21-30	31 or more
1995	16.9	13.4	20.9	13.4	11.3	6.2	17.9
2006	14.3	11.6	18.5	12.9	11.5	6.8	24.3
% change from 1996	-15.4	-13.4	-11.5	-3.8	+1.8	+9.6	+35.8

Table 5: DAFW Cases by Duration: 1995-2006
Note. Data from "Lost Worktime Injuries and Illnesses: Characteristics and Resulting Time Away From Work, 1995-2006," by Bureau of Labor Statistics, 2007. Washington, DC: U.S. Department of Labor, Author. DAFW = days away from work.

resulting in lost work-time and the DART rate (which includes cases with days away from work, job transfer or restriction) have reduced substantially. From 2000 to 2006, the number of lost work-time cases dropped by 480,518 (28.9%), while the DART dropped 23.3%. Those reductions are commendable.

BLS data on lost work-time injuries and illnesses tracks well with NCCI reports with respect to the lost workday categories in which the reductions occurred. Data in Table 5 are from the BLS's *Lost Work-Time Injuries and Illnesses: Characteristics and Resulting Time Away From Work* reports for the years 1995 and 2006. Table 10 in those reports is titled "Percent distribution of nonfatal occupational injuries and illnesses involving days away from work." It shows the percentages of select days-away-from-work categories as each category relates to the total number of days-away-from-work cases reported in a given year.

The decreases (the trends) in the percentages for the first four days-away-from-work categories are noteworthy. The frequency of incidents resulting in lesser injury is down. For the 11-to-20-days-away category, the increase of 1.8% only begins to show an upward trend. The 9.6% increase for the 21-to-30- days-away category deserves attention, as does the increase of 35.3% for the 31 or more days-away category. Given this, it is recommended that safety professionals determine whether increases in the 21-to 30 and the 31-or-more days-away categories have occurred in their organizations' operations.

In 2002, OSHA revised the rules on how days away from work are counted, so the trend data in Table 5 need a closer look. Using the base data from the BLS reports for the years 1995 through 2001, and assuming the rules had not changed, Alan Hoskin, a statistician formerly with National Safety Council, statistically projected numbers for the years 2002 and 2003. He found that the differences are small—1.2% and 1.7%—and do not greatly affect the trend data.

One cannot conclude from the BLS data that the number of incidents resulting in severity of 21 to 30 or 31 or more days away from work has increased. The data in Table 4 show that the number of lost work-time cases has been significantly reduced. The data in Table 5 indicate that incidents resulting in severity are a larger segment of all days-away-from-work cases reported and that serious injuries have not been reduced at the same rate as less severe injuries.

Debunking the Myth

To further reduce serious injuries and fatalities, safety professionals must address a long-held and still applied belief that reducing incident frequency will equivalently reduce incidents that result in severity. The data in this article convincingly show that this premise is unsustainable.

Others have raised the issue as well. At the 2003 Behavioral Safety Now Conference, James Johnson, a managing director at Liberty Mutual Insurance Co., said:

> I'm sure that many of us have said at one time or another that frequency reduction will result in severity reduction. This popularly held belief is not necessarily true. If we do nothing different than we are doing today, these types of trends will continue.

In 2004, DNV Consulting issued an invitation to the process industry titled "Leading Indicators for Major Accident Hazards: An Invitation to Industry Partners." The goal was to get the industry to finance research into the causal factors for major accidents—a goal that was not achieved. In part, this invitation stated:

> Much has been said about the classical loss control pyramid, which indicates the ratio between no loss incidents, minor incidents and major incidents, and it has often been argued that if you look after the small incidents, the major loss incidents will improve

also. The major reality, however, is somewhat different. If you manage the small accidents effectively, the small accident improves, but the major accident rate stays the same, or even slightly increases.

To recognize that the premises on which the pyramids, the triangles or the specific ratios (e.g., the 300-29-1 ratios) were built are not valid requires a major concept change—and the data show this is necessary.

Consider, also, the symmetry between what Johnson said in 2003 and a philosophical statement of Yogi Berra: "If you keep doing what you did, you will keep getting what you got." Listen to these voices and those of DNV and this author. It is obvious that frequency reduction does not necessarily produce equivalent severity reduction. If safety professionals propose nothing different with respect to safety management systems than they have proposed in the past, serious injury potential will not be significantly reduced. The data require that safety professionals adopt a new mindset—one that results in a targeted focus on preventing low-probability/serious-consequence events.

Characteristics of Incidents Resulting in Severe Injuries and Fatalities

As safety professionals study the characteristics of incidents that result in serious injuries and fatalities to select predictive indicators from those data, they should consider the following general observations based on the author's analyses of more than 1,200 incident investigation reports.

1) A large proportion of incidents resulting in serious injuries and fatalities occur:

a) when unusual and nonroutine work is being performed;

b) when upsets occur—meaning normal operations become abnormal;

c) in nonproduction activities;

d) where sources of high energy are present;

e) in what can be called at-plant construction operations (e.g., a motor that weighs 800 lb and sits on a platform 15 ft above the floor needs to be replaced, and the work will be performed by in-house personnel).

2) Many incidents resulting in serious injuries and fatalities are unique and singular events, having multiple and complex causal factors that may have organizational, technical, operational systems or cultural origins.

3) Causal factors for low-probability/high-consequence events are not represented in the analytical data on incidents that occur frequently and result in minor injury. However, such incidents, occurring in routine work, may be predictors of severity

potential if a high energy source was present (e.g., operation of powered mobile equipment, electrical contacts). Also, certain ergonomics-related incidents are exceptions.

4) The quality of the incident investigation reports reviewed was, on average, abysmal. A large percentage of the investigations stopped when human error—the so-called unsafe act—was identified and the corrective action focused on modifying worker behavior. The investigations seldom proceeded upward into the decision making that may have influenced what the worker did.

Guidelines for Preventing Human Error in Process Safety, published by the Center for Chemical Process Safety (1994), contains two chapters that provide a primer on human error reduction. Excerpts from that text follow.

> It is readily acknowledged that human errors at the operational level are a primary contributor to the failure of systems. It is often not recognized, however, that these errors frequently arise from failures at the management, design or technical expert levels of the company.
>
> One of the central principles in this book is the need to consider the organizational factors that create the preconditions for errors, as well as the immediate causes.

Specifics From Certain Studies

Supporting the foregoing general observations, the following specifics were noted in the experience of individual companies when analyses were made of serious injuries and fatalities.

- Thirty-five percent of serious injuries and fatalities were triggered by a deviation from normal operations (upsets).
- Over a 10-year period, 51% of fatalities occurred to contractor employees.
- In three companies with a combined total of 230,000 employees, each having low OSHA incidence rates, composite data indicated that 74% of lost workday cases with days away from work involved ancillary and support personnel.
- For companies with incidence rates higher than their industry's average, and in companies where the work involves heavy materials handling or is highly repetitive, the percentage of severe injuries occurring to production personnel was higher than for those to support personnel.
- About 50% of major accidents involved the operation of powered mobile equipment (e.g., forklifts, cranes).
- Reviews of serious injuries and fatalities involving exposure to electric current indicate that while lockout/tagout systems may have met OSHA and National Electrical Code requirements, the design of the systems produced error-inducing situations (e.g., lockout stations were not conveniently located).

•Hazards and risks were not adequately addressed during the design process, and inadequate design features often appeared as causal factors in incident investigation reports.

•Having effective management of change procedures in place would have greatly reduced major accident potential.

Petersen (1998) also subscribed to the view that serious injury and fatality potential need special attention.

> If we study any mass data, we can readily see that the types of accidents that result in temporary total disabilities are different from the types of accidents resulting in permanent partial disabilities or in permanent total disabilities or fatalities. The causes are different. There are different sets of circumstances surrounding severity. Thus, if we want to control serious injuries, we should try to predict where they will happen.

Since studies have established that the causal factors and the circumstances surrounding incidents which result in serious injuries are different, safety professionals should try to predict where serious injuries and fatalities may occur, and recommend improvements necessary in the relative safety management systems so as to avoid their occurrence.

Significance of Organizational Culture

Since causal factors for incidents resulting in serious injuries and fatalities are largely systemic and a reflection of the organization's safety culture, that subject must be explored. Comments from the *Report of the Columbia Accident Investigation Board* (NASA, 2003) are pertinent.

> The physical cause of the loss of Columbia and its crew was a breach in the thermal protection system on the leading edge of the left wing. In our view, the NASA organizational culture had as much to do with this accident as the foam.

In every organization, its culture—values, norms, beliefs, myths and practices—is translated into a system of expected behavior. That expected behavior positively or negatively impacts decisions made with respect to management systems, design and engineering, operating methods, work methods and prescribed task performance.

For many workplace incidents that result in serious consequences there has been, over time, a continuum of less-than-adequate safety decisions that created a system of expected behavior which condoned considerable risk taking. Management decisions shape the corporate culture and create error-producing factors.

Reason (1997) also discusses the accumulation of systemic causal factors.

> Latent conditions, such as poor design, gaps in supervision, undetected manufacturing defects or maintenance failures, unworkable procedures, clumsy automation, shortfalls

in training, less than adequate tools and equipment, may be present for many years before they combine with local circumstances and active failures to penetrate the system's layers of defenses.

They arise from strategic and other top-level decisions made by governments, regulators, manufacturers, designers and organizational managers. *The impact of these decisions spreads throughout the organization, shaping a distinctive corporate culture and creating error-producing factors within the individual workplaces* (emphasis added).

The Current Business Climate: Effect on Organizational Culture and Decision Making

Both a literature review and discussions with safety professionals require that consideration be given to the current economic climate and its possible effect on an organization's safety culture. Consider the following statements from the *Report of the OECD Workshop on Lessons Learned from Chemical Accidents and Incidents.* (OECD is the Organization for Economic Cooperation and Development, an international group.)

> The concept of "drift" as defined by [Jens] Rasmussen was generally agreed upon as being far too common in the current business environment. Rasmussen defined "drift" as the systematic organizational performance deteriorating under competitive pressure, resulting in operation outside the design envelope where preconditions for safe operation are being systematically violated (OECD, 2005).

The OECD report also includes comments attributed to Norika Hama, a professor of international economics at Doshisha University Business School:

> In their bid to make profit under deflationary pressures, [Japanese] companies have been restructuring their operations and trying to cut costs, and are compelled to continue using facilities and equipment that normally would have been replaced and renewed years ago, thereby raising the risk of accidents. Also because of job cuts, the firms do not have sufficient numbers of workers who can repair and keep the old equipment in proper condition.
>
> The operation of Japan's manufacturing industries was once looked upon as a global standard, but the fact that major companies that are supposed to symbolize that standard have been hit by serious accidents shows deflation has damaged the nation's industrial base.

Also consider what Rasmussen (1997) says about risk management:

> Companies today live in a very aggressive and competitive environment which will focus the incentives of decision makers on short-term financial and survival criteria rather than long-term criteria concerning welfare, safety and the environment. Studies of several accidents revealed that they were the effects of a systematic migration of organizational behavior toward accident under the influence of pressure toward cost-effectiveness in an aggressive, competitive environment.

Comments from the U.S. Chemical Safety and Hazard Investigation Board (CSB) report on the 2005 BP Texas City, TX, explosion that resulted in 15 deaths and 180 injuries are also pertinent.

> The Texas City disaster was caused by organizational and safety deficiencies at all levels of the BP Corp. Warning signs of a possible disaster were present for several years, but company officials did not intervene effectively to prevent it. Cost cutting and failure to invest left the Texas City refinery vulnerable to a catastrophe. BP targeted budgeted cuts of 25% in 1999 and another 25% in 2005, even though much of the refinery's infrastructure and process equipment were in disrepair.

In a March 20, 2007, CSB news release, then chair Carolyn Merritt said, "The combination of cost-cutting, production pressures and failure to invest caused a progressive deterioration of safety at the refinery." The impact of economics on decisions that may have a negative effect on the safety culture must be taken seriously.

Assume that an organization's senior executives want to know about the economics-related predictors for serious injury potential that may exist in their operations and that safety professionals want to conduct a study to identify them. Such a study can be built on the following outline.

1) In the current business climate, does the incentive system for decision makers result in focusing on short-term financial goals, resulting in drift—the systematic organizational performance deteriorating under competitive pressure?

2) Has the gap widened between issued policy and procedure, and what actually takes place at the company's locations?

3) Does the organization continue using facilities and equipment that normally would have been replaced years ago, thereby raising the risk of serious injuries and fatalities?

4) Has there been a high turnover of location managers, the result being considerable variation in the emphasis on safety management?

5) Is staffing at all levels sufficient—both as to number and qualifications—to maintain a superior level of safety performance?

6) Because of staff cuts, does the firm have insufficient numbers of qualified workers who can repair and keep equipment in proper condition?

7) Has complacency developed at the site due to presumed superior performance, as measured by OSHA statistics?

8) Do safety audits lack the depth needed to identify continuing deterioration in management systems that results in greater risk?

Every element in this list relates to concerns expressed by safety professionals about deterioration in safety management systems as they comment on trends in their organizations.

Avoiding Self-Delusion

With respect to the Texas City incident, CSB (2005) also says that "a very low personal injury rate at Texas City gave BP a misleading indicator of process safety performance." Others have similarly become aware that low injury incidence rates have little predictive value for severity potential. In a speech at the International Association of Oil and Gas Producers Offshore Safety Forum, Volkert Zijlker (2005), chair of the Oil and Gas Producers Safety Committee, said:

> We need to differentiate our focus on recurring safety incidents commensurate to the escalation potential. We concluded that TRIR/LTIF have little predictive value toward the potential escalation to single and multiple fatalities. They also tell us little about major accident risk.

Neither safety professionals nor executive management should delude themselves into believing that achieving low OSHA incidence rates ensures that serious injuries will not occur.

Actions to Reduce Serious Injury Potential

With a concentrated focus on further preventing serious injuries, safety professionals should consider the following initiatives:

1) Propose a study of serious injuries and fatalities in the entities to which they give counsel.

2) Significantly improve the quality of incident investigations.

3) Conduct a gap analysis, emphasizing the prevention through design provisions in ANSI/AIHA Z10-2005.

4) Initiate a system such as the critical incident technique (NSC, 2001a; Infopolis 2 Consortium) to gather information on near-hits.

Propose a Study of Serious Injuries and Fatalities

To produce information that relates directly to the entities to which safety professionals give counsel, it is proposed that serious injuries and fatalities which have occurred in those entities be studied. Such studies will not be time-consuming since the data to be collected and analyzed should already exist or be easily obtained. A model instrument that can be used in this study is shown in Figure 1; it should be modified to suit an organization's structure, culture, inherent risks, operations specifics and incident experience.

The study should seek predictive indicators, represented by shortcomings in safety management systems, so that improvement can be proposed. Item 8 in the survey instrument pertains to causal factors and would address those pertinent to the operations being studied.

1. Enter incident descriptive data_____		
2. Enter NAICS code_____	Job title(s)_____	
3. Type of event	Serious Injury	3-A____
	Fatality	3-B____
	Near hit—life threatening	3-C____
4. Type of work code	Main operation/principle business	4-A____
	Operation support/ancillary personnel	4-B____
	Motor vehicle, other than sales	4-C____
	Construction/renovation	4-D____
	Sales	4-E____
	Office	4-F____
5. Work/task performed	Often/continuously/routine	5-A____
	Only occasionally	5-B____
	Seldom/quarterly or less	5-C____
6. Enter	Employee	6-A____
	Contractor	6-B____
7. Injury type	Fall	7-A____
	Struck by/against	7-B____
	Electrocution	7-C____
	Caught in/between	7-D____
	Motor vehicle (auto/truck)	7-E____
	Other mobile equipment	7-F____
	Exposure to chemical substance	7-G____
	Other	7-H____
8. Identify and enter the causal factors (design and engineering, operational, system, cultural, organizational, etc.)		____

Figure 1: Serious Injury and Fatality Review Instrument

Improve Incident Investigations

While the reality of the design and engineering, operational systems and cultural causal factors should be identified and analyzed in the proposed study, safety professionals should not be surprised to find that the incident investigation reports lack in-depth causal factors determination. As noted, the author's studies of 1,200 reports have found that incident investigations seldom reveal the core causal factors.

Comments by the Columbia Accident Investigation Board (NASA, 2003) match the conclusions drawn by this author through his research. While reading the following excerpts from that group's report, safety professionals should think about how they relate to the quality of the incident investigation systems in their organizations.

> Many accident investigations do not go far enough. They identify the technical cause of the accident, and then connect it to a variant of "operator error." But this is seldom the entire issue. When the determinations of the causal chain are limited to the technical flaw and individual failure, typically the actions taken to prevent a similar event in the future are also limited: fix the technical problem and replace or retrain the individual responsible. Putting these corrections in place leads to another mistake—the belief that the problem is solved.
>
> Too often, accident investigations blame a failure only on the last step in a complex process, when a more comprehensive understanding of that process could reveal that earlier steps might be equally or even more culpable.
>
> In this board's opinion, unless the technical, organizational and cultural recommendations made in this report are implemented, little will have been accomplished to lessen the chance that another accident will follow.

As noted, many incidents resulting in serious injuries are unique and singular events, having multiple and complex causal factors that may have organizational, technical, operational systems or cultural origins. Substantial reductions in serious injuries are unlikely if incident investigation systems are not improved to address the reality of their causal factors.

The 5 Why System

One way to improve an incident investigation system is to use the 5 why technique. Highly skilled incident investigators may say that this technique is inadequate because it does not promote the identification of root causal factors that result from decisions made at a senior executive level. Nevertheless, achieving competence in applying this concept will be a major step forward in many organizations.

The origin of the 5 why process is attributed to Taiichi Ohno. While he was at Toyota, Ohno developed and promoted a practice of asking why five times to determine what caused a problem so that root causal factors can be identified and effective countermeasures implemented. The process is applied in a large number of settings for a wide range of problems.

Since the premise on which the 5 why concept is based is uncomplicated, it can be (and has been) easily incorporated into the incident investigation process. For more complex incident situations, starting with the 5 why strategy may lead to the use of event trees or fishbone diagrams or more sophisticated investigation systems (iSixSigma).

Conduct a Gap Analysis

Approval of ANSI/AIHA Z10-2005 was a major development. Provisions in Z10 are state of the art. To identify shortcomings in safety management systems that relate particularly to serious injury prevention, it is suggested that safety professionals conduct a gap analysis to compare existing safety management systems to the provisions in Z10.

While this analysis should include all provisions in the standard, the focus here is on prevention through design processes since most companies will find shortcomings in their safety management systems concerning them (Manuele, 2008). Improvements in these processes should reduce serious injury potential.

•**Design reviews**. Z10 requires that processes be in place to conduct safety-related design reviews so as to avoid bringing hazards and risks into the workplace.

•**Risk assessments**. Hazards are to be identified and analyzed, and risks are to be assessed and prioritized.

•**Hierarchy of controls**. An organization must implement and maintain a process for achieving feasible risk reduction based on a prescribed hierarchy of controls.

•**Management of change**. The objective of a management of change system is to prevent introducing hazards and risks into the work environment when operational changes are made. Given the author's studies of incident experience—in which it was noted that many incidents resulting in serious injuries occur when unusual work is done (e.g., as when changes are made)—safety professionals should strongly consider proposing the adoption of such a system.

•**Procurement**. Z10 requires that safety specifications be included in purchasing and acquisition processes to avoid bringing hazards and risks into the workplace.

As the gap analysis proceeds, the system shortcomings identified should be evaluated with respect to their being predictive of the probability that major incidents may occur.

Encourage Use of a Variation of the Critical Incident Technique

The proposed survey instrument (Figure 1) contains provisions to enter data on life-threatening near-hits. Safety professionals should consider adopting a system—such as the critical incident technique (NSC, 2001; Infopolis 2 Consortium)—to collect data on near-hits and out-of-the-norm situations to capture the predictive value such data provide. The purpose of applying the technique is to identify and address hazards that have serious injury potential.

A system requiring interviews, form completion or computer entry is created whereby employees are asked for their input on serious injury potential, including near-hit hazardous situations. For the process to succeed, one must recognize that workers are a valuable resource in identifying hazards and risks because of their extensive knowledge of how the work gets done.

With respect to incident recall, Johnson (1980) says:

> Such [incident recall] studies, whether by interview or questionnaire, have a proven capacity to generate a greater quantity of relevant, useful reports than other monitoring techniques, so much so as to suggest that their presence is an indispensable criterion of an excellent safety program.

A system that seeks to identify causal factors before their potentials are realized would serve well in attempts to avoid low-probability/serious-consequence events.

Conclusion

To reduce the potential for major accidents, management must embed that purpose within its culture. This will ensure that avoiding the causal factors for severe injuries is considered in the application of every element in the safety management system.

Achieving this requires a new mindset—in every aspect of safety management, from the design process to dismantling and disposition—and giving serious injury prevention a higher priority. The intent would be to achieve an understanding that personnel at all levels have a particular responsibility to:

- anticipate, predict and take corrective action on hazards and risks that may have serious injury or fatality potential;
- ensure that root causal factors for incidents which result in severe injuries are reviewed in depth;
- identify predictive indicators, including knowledge obtained from studies of near-hits;
- address organizational, operational, technical and cultural causal factors.

As safety professionals study serious injury causal factors and identify the improvements needed in safety management systems, they may find that a culture change is necessary. This would require them to take a significant leadership role. ∎

References

Alcoa. (2007, Jan. 22). Alcoa Foundation awards $100,000 grant to Indiana University of Pennsylvania to support national safety forum [News release]. Pittsburgh, PA: Author.

Bureau of Labor Statistics (BLS). (2007). Census of fatal occupational injuries, 1997-2006. Washington, DC: U.S. Department of Labor, Author.

BLS. (2007). Lost worktime injuries and illnesses: Characteristics and resulting time away from work, 1995-2006. Washington, DC: U.S. Department of Labor, Author.

Center for Chemical Process Safety (CCPS). (1994). *Guidelines for preventing human error in process safety*. New York: American Institute of Chemical Engineers, Author.

Chemical Safety and Hazard Investigation Board (CSB). (2007, March 20). U.S. Chemical Safety Board investigators conclude "organizational and safety deficiencies at all levels of the BP Corporation" caused March 2005 Texas City disaster that killed 15, injured 180 [News release]. Washington, DC: Author.

CSB. (2007, March). *Investigation report: Refinery explosion and fire (15 killed, 180 injured), BP, Texas City, TX, March 23, 2005* (Report No. 2005-04-I-TX). Washington, DC: Author. Retrieved Oct. 24, 2008, from http://www.csb.gov/completed_investigations/docs/CSBFinalReportBP.pdf.

Infopolis 2 Consortium. Task analysis methods: Critical incident technique. Aix-en-Provence, France: Author. Retrieved Oct. 24, 2008, from http://www.ul.ie/~infopolis/methods/incident.html.

iSixSigma. Determine the root cause: 5 whys. Retrieved Oct. 24, 2008, from http://www.isixsigma.com/library/content/c020610a.asp.

Johnson, J. (2003). Unpublished speech at the 2003 Behavioral Safety Now Conference, Reno, NV, USA.

Johnson, W.G. (1980). *MORT safety assurance systems.* New York: Marcel Dekker.

Manuele, F.A. (2008). *Advanced safety management: Focusing on Z10 and serious injury prevention.* Hoboken, NJ: John Wiley & Sons.

Mealy, D. (2005). *State of the line report.* Boca Raton, FL: National Council on Compensation Insurance (NCCI).

NASA. (2003, Aug.). *Columbia accident investigation report: Volume 1.* Washington, DC: Author. Retrieved Oct. 24, 2008, from http://www.nasa.gov/columbia/home/CAIB_Vol1.html.

National Safety Council (NSC). (1995). *Accident facts.* Itasca, IL: Author.

NSC. (2001a). *Accident prevention manual for industrial operations* (12th ed.). Itasca, IL: Author.

NSC. (2001b). *Injury facts.* Itasca, IL: Author.

National Council on Compensation Insurance (NCCI). (2006, Nov.). *The remarkable story of declining frequency—down 30% in the past decade* [video]. Boca Raton, FL: Author. Retrieved Oct. 24, 2008, from https://www.ncci.com/nccimain/IndustryInformation/NCCIVideos/ArchivedArticles/Pages/video_declining_frequency_11-06.aspx.

NCCI. (2005, June). Workers' compensation claim frequency down again [Research brief]. Boca Raton, FL: Author.

Organization for Economic Cooperation and Development (OECD). (2005). Report of the OECD workshop on lessons learned from chemical accidents and incidents. Sept. 21-23, 2004, Karlskoga, Sweden. Paris, France: Author. Retrieved Oct. 24, 2008, from http://appli1.oecd.org/olis/2005doc.nsf/43bb6130e5e86e5fc12569fa005d004c/e6de9b632c0d0368c1256fd400348fa5/$FILE/JT00182564.PDF.

Petersen, D. (1998). *Safety management* (2nd ed.). Des Plaines, IL: ASSE.

Rasmussen, J. (1997). Risk management in a dynamic society: A modelling problem. *Safety Science, 27*(2/3), 183-213.

Reason, J. (1997). *Managing the risks of organizational accidents.* Burlington, VT: Ashgate Publishing Co.

Zijlker, V. (2005). What are the major health, safety and regulatory issues and concerns in worldwide operations? Paper presented at International Regulators Forum Offshore Safety Forum, March 31-April 1, 2005.

8 ACCEPTABLE RISK
TIME FOR SH&E PROFESSIONALS TO ADOPT THE CONCEPT

ORIGINALLY PUBLISHED MAY 2010

The term *acceptable risk* is frequently used in standards and guidelines throughout the world, yet a substantial percentage of those with SH&E responsibilities are reluctant to adopt or use it. Evidence of this reluctance often arises in discussions surrounding the development of new or revised standards or technical reports. The aversion may derive from:

•a lack of awareness of the nature of risk;

•concern over the subjective judgments made and the uncertainties that almost always exist when risks are assessed;

•the lack of in-depth statistical probability and severity data that allows precise and numerically accurate risk assessments;

•insufficient real-world experience in more hazardous environments where non-trivial risks are necessarily accepted every day.

However, in recent years, the concept of acceptable risk has been interwoven into international standards and guidelines for a broad range of equipment, products, processes and systems. This has occurred in recognition of the fact that risk-related decisions are made constantly in real-world applications and that society benefits if those decisions achieve acceptable risk levels.

This primer is designed to help readers gain an understanding of risk and the concept of acceptable risk. The far-reaching premise presented is fundamental in dealing with risk. Several examples of the use of the term acceptable risk as taken from the applicable literature. Discussions address the impossibility of achieving zero risk levels, the inadequacy of *minimum risk* as a replacement term for acceptable risk, and the shortcomings that may result from designing only to a standard's requirements. Finally, the "as low as reasonably practicable (ALARP) concept" is presented with an example of how it is applied in achieving an acceptable risk level.

Fundamental Premise

The following general, all-encompassing premise is basic to the work of all personnel who give counsel to prevent injury, illness and damage to property and the environment.

> The entirety of purpose of those responsible for safety, regardless of their titles, is to identify, evaluate, and eliminate or control hazards so that the risks deriving from those hazards are acceptable.

That premise is supported by this theory: If there are no hazards, if there is no potential for harm, risks of injury or damage cannot arise. If there were no risks, there would be no need for SH&E professionals. (Note: For simplicity, the terms hazard, risk and safety apply to all hazard-related incidents or exposures that could cause injury or illness, or damage property or the environment.)

Use of the Term Acceptable Risk

The more frequent use over time of the term acceptable risk in standards and guidelines is notable, as the following citations show. SH&E personnel reluctant to adopt the concept implied by the term should consider the breadth and implication of this evolution. The term acceptable risk is becoming the norm. The following (intentionally lengthy) list of citations shows how broadly the concept of acceptable risk has been adopted.

1) Lowrance (1976) wrote, "A thing is safe if its risks are judged to be acceptable."

2) The following citation, from a 1980 court decision, is significant because it has given long-term guidance with respect to Department of Labor policy and to the work performed by NIOSH.

The Supreme Court's benzene decision of 1980 states that "before he can promulgate any permanent health or safety standard, the Secretary [of Labor] is required to make a threshold finding that a place of employment is unsafe—in the sense that significant risks are present and can be eliminated or lessened by a change in practices" (*Industrial Union Department, AFL-CIO v. American Petroleum Institute U.S. at 642*). The Court broadly describes the range of risks OSHA might determine to be significant: It is the agency's responsibility to determine in the first instance what it considers to be a "significant" risk. *Some risks are plainly acceptable and others are plainly unacceptable* (emphasis added).

For example, if the odds are 1 in 1 billion that a person will die from cancer by taking a drink of chlorinated water, the risk clearly could not be considered significant. On the other hand, if the odds are 1 in 1,000 that regular inhalation of gasoline vapors that are 2% benzene will be fatal, a reasonable person might consider the risk significant and take appropriate steps to decrease or eliminate it. The Court further stated:

> The requirement that a "significant" risk be identified is not a mathematical straitjacket. Although the agency has no duty to calculate the exact probability of harm, it does have an obligation to find that a significant risk is present before it can characterize a place of employment as "unsafe" and proceed to promulgate a regulation.

3) International Organization for Standardization (ISO) and International Electrotechnical Commission (IEC) (1990) issued guidelines for including safety aspects in standards. These guidelines provide standardized terms and definitions to be used in standards for "any safety aspect related to people, property or the environment." The second edition, issued in 1999, contains the following definitions:

Safety: freedom from unacceptable risk (3.1).

Tolerable risk: risk which is accepted in a given context based on the current values of society (3.7).

4) Fewtrell and Bartram (2001), in a document for World Health Organization, address standards related to water quality. They offer the following guidelines for determining acceptable risk.

> A risk is acceptable when: it falls below an arbitrary defined probability; it falls below some level that is already tolerated; it falls below an arbitrary defined attributable fraction of total disease burden in the community; the cost of reducing the risk would exceed the costs saved; the cost of reducing the risk would exceed the costs saved when the "costs of suffering" are also factored in; the opportunity costs would be better spent on other, more pressing, public health problems; public health professionals say it is acceptable; the general public say it is acceptable (or more likely, do not say it is not); politicians say it is acceptable.

5) OSHA (2003) set forth requirements for organizations seeking certification under the agency's Voluntary Protection Programs (VPP):

> **Worksite Analysis.** A hazard identification and analysis system must be implemented to systematically identify basic and unforeseen safety and health hazards, evaluate their risks, and prioritize and recommend methods to eliminate or control hazards to an acceptable level of risk.

6) ANSI/ASSE Z244.1-2003(R2009) on lockout/tagout states, "A.2: Acceptable level of risk: If the evaluation in A.1.6 determines the risk to be acceptable, then the process is completed. . . ."

7) UN (2009) offers this definition when addressing basic terms of disaster risk reduction: "Acceptable risk: The level of potential losses that a society or community considers acceptable given existing social, economic, political, cultural, technical and environmental conditions."

8) The online Sci-Tech Dictionary (accessed at www.answers.com/topic/acceptable-risk-geophysics) provides this definition of acceptable risk as the term is used in geology:

> **Acceptable risk:** (geophysics) In seismology, that level of earthquake effects which is judged to be of sufficiently low social and economic consequence, and which is useful for determining design requirements in structures or for taking certain actions.

9) Australia/New Zealand AS/NZS 4360: 2004 risk management standard uses this definition (in 1.3.16): "Risk acceptance: An informed decision to accept the consequences and the likelihood of a particular risk."

10) ANSI/AIHA Z10-2005 contains the following citations with respect to acceptable risk.

> E5.1.1: Often, a combination of controls is most effective. In cases where the higher order of controls (elimination, substitution and implementation of engineering controls) does not reduce the risk to an acceptable level, lower order controls may be necessary.
>
> Appendix E (Informative), Assessment and Prioritization (Z10 Section 4.2): The last sentence in Step 7 in a Hazard Analysis and Risk Assessment Guide says: "The organization must then determine if the level of risk is acceptable or unacceptable."

A definition of residual risk follows the hazard analysis and risk assessment guide in Z10:

> Risk can never be eliminated entirely, though it can be substantially reduced through application of the hierarchy of controls. Residual risk is defined as the remaining risk after controls have been implemented. It is the organization's responsibility to determine whether the residual risk is acceptable for each task and associated hazard. Where the residual risk is not acceptable, further actions must be taken to reduce risk.

11) DOT's Pipeline and Hazardous Materials Safety Administration (PHMSA, 2005) has issued risk management definitions, including this one:

> **Acceptable risk:** An acceptable level of risk for regulations and special permits is established by consideration of risk, cost/benefit and public comments. Relative or comparative risk analysis is most often used where quantitative risk analysis is not practical or justified. Public participation is important in a risk analysis process, not only for enhancing the public's understanding of the risks associated with hazardous materials transportation, but also for ensuring that the point of view of all major segments of the population-at-risk is included in the analyses process.

Risk and cost/benefit analysis are important tools in informing the public about the actual risk and cost as opposed to the perceived risk and cost involved in an activity. Through such a public process PHMSA establishes hazard classification, hazard communication, packaging and operational control standards.

12) ANSI/PMMI B155.1-2006 on packaging machinery and packaging-related converting machinery contains this definition: "Acceptable risk: risk that is accepted for a given task or hazard. For the purpose of this standard the terms *acceptable risk* and *tolerable risk* are considered synonymous (3.1)."

13) In the 2007 revision of BS OHSAS 18001:2007 on occupational health and safety management systems, British Standards Institution (BSI) made a significant change. Specifically, the term *tolerable risk* was replaced with the term *acceptable risk* (3.1).

14) In the introduction of IEC 60601-1-9 (2007), which addresses medical equipment design, IEC states, "The standard includes the evaluation of whether risks are acceptable (risk evaluation)."

15) A machinery safety document issued in 2009 by the Institute for Research for Safety and Security at Work and the Commission for Safety and Security at Work in Quebec, Canada, states, "When machine-related hazards . . . cannot be eliminated through inherently safe design, they must then be reduced to an acceptable level."

16) ASSE's (2009) technical report on prevention through design includes the following information:

Scope and Purpose
1.3 The goals of applying prevention through design concepts are to:
1.3.1 Achieve that state for which risks are at an acceptable level.

Definitions
Acceptable risk: That risk for which the probability of a hazard-related incident or exposure occurring and the severity of harm or damage that may result are as low as reasonably practicable (ALARP) and tolerable in the setting being considered.
ALARP: that level of risk which can be further lowered only by an increment in resource expenditure that is disproportionate in relation to the resulting decrement of risk.

About the Foregoing Citations

1) Since it is almost always the case that resources are limited, this phrase in the WHO citation, "the opportunity costs would be better spent on other, more pressing problems," has a significant bearing on risk acceptance decision making and on priority setting.

2) Several citations relate to the fact that residual risk cannot be eliminated entirely and that residual risk acceptance decisions are commonly and frequently made.

Whenever a production machine is turned on, a residual risk level is being accepted. Every time a design decision is made or a product design is approved, those making the decision approve a residual and acceptable risk level.

3) Definitions of acceptable risk nearly identical to that in ANSI/PMMI B155.1-2006 appear in ANSI B11-2008, General Safety Requirements Common to ANSI B11 Machines, and ANSI/AMT B11.TR7-2007, ANSI Technical Report for Machines: A Guide on Integrating Safety and Lean Manufacturing Principles in the Use of Machinery.

4) Replacing the term *tolerable risk* with *acceptable risk* in BS OHSAS 18001 by an organization as influential as BSI is noteworthy. In some parts of the world, because of requirements in contract bid situations, companies must show that their safety management systems are "certified." BS OHSAS 18001 is often the basis of such certification. This modification by BSI indicates that the goal to be achieved is acceptable risk levels.

As the cited references illustrate, the concept of acceptable risk has been broadly adopted internationally, and the term is becoming the norm. SH&E professionals who are reluctant to adopt this concept would do well to recognize that they have an obligation to be current with respect to the state of the art and reconsider their views.

The Nature and Source of Risk

Risk is expressed as an estimate of the probability of a hazard-related incident or exposure occurring and the severity of harm or damage that could result. All risks with which SH&E professionals deal derive from hazards without exception. A hazard is defined as the potential for harm. Hazards include all aspects of technology and activity that produce risk. Hazards include the characteristics of things (e.g., equipment, dusts, chemicals) and the actions or inactions of people.

The probability aspect of risk is defined as the likelihood of an incident or exposure occurring that could result in harm or damage—for a selected unit of time, events, population, items or activity being considered. The severity aspect of risk is defined as the degree of harm or damage that could reasonably result from a hazard-related incident or exposure.

Comparable statements and definitions appear in much of the current literature on risk and acceptable risk. One resource, the *Framework for Environmental Health Risk Management* (The Presidential Congressional Commission on Risk Assessment and Risk Management, 1997), was selected for citation because of its broad implications. Excerpts follow.

What Is "Risk"

Risk is defined as the probability that a substance or situation will produce harm under specified conditions. Risk is a combination of two factors:
- the probability that an adverse event will occur;
- the consequences of the adverse event.

Risk encompasses impacts on public health and on the environment, and arises from exposure and hazard. Risk does not exist if exposure to a harmful substance or situation does not or will not occur. Hazard is determined by whether a particular substance or situation has the potential to cause harmful effects. Risk . . . is the probability of a specific outcome, generally adverse, given a particular set of conditions.

Residual risk . . . is the health risk remaining after risk reduction actions are implemented, such as risks associated with sources of air pollution that remain after implementation of maximum achievable control technology.

Risk assessment . . . is an organized process used to describe and estimate the likelihood of adverse health outcomes from environmental exposures to chemicals. The four steps are hazard identification, dose-response assessment, exposure assessment and risk characterization.

Zero Risk: Not Attainable

It has long been recognized that zero risk levels are not attainable. If a facility exists or an activity proceeds, it is impossible to realistically conceive of a situation that presents no probability of an adverse incident or exposure occurring. According to Lowrance (1976):

> Nothing can be absolutely free of risk. One can't think of anything that isn't, under some circumstances, able to cause harm. Because nothing can be absolutely free of risk, nothing can be said to be absolutely safe. There are degrees of risk and, consequently, there are degrees of safety.

Similar comments appear in ISO/IEC Guide 51, under "The Concept of Safety" (section 5):

> There can be no absolute safety: some risk will remain, defined in this guide as residual risk. Therefore a product, process or service can only be relatively safe. Safety is achieved by reducing risk to a tolerable level, defined in this guide as tolerable risk.

In the real world, attaining a zero risk level, whether in the design or redesign processes or in facility operations, is not possible. That said, after risk avoidance, elimination or control measures are taken, the residual risk should be acceptable, as judged by the decision makers.

Also, one must recognize that inherent risks which are acceptable and tolerable in some occupations are not tolerable in others. For example, some work conditions considered tolerable in deep sea fishing (e.g., a pitching and rolling work floor, the ship's deck) would not be tolerable in other work settings. In other situations, such as for certain chemical or radiation exposures designed to function at higher than commonly accepted permissible exposure levels, the residual risk will be judged as unacceptable and operations at those levels would not be permitted.

Nevertheless, society accepts continuation of certain operations with high occupational and environmental risks. This is demonstrated by fatality rate data from the Bureau of Labor Statistics (Table 1). The fatality rate (rounded) is the rate per 100,000 workers. The national average fatality rate for all private industries is 4.0.

Although the fatality rates among all employment categories are highest for the occupations highlighted in Table 1, the public has not demanded that the operations in which they occur cease. The inherent risks in the high-hazard categories are considered *tolerable*. It should be recognized that considerable research has been undertaken to make those occupations safer.

Occupation	Fatality rate[a]
Fishers and related fishing workers	112
Logging workers	86
Aircraft pilots and flight engineers	67
Structural iron and steel workers	46
Farmers and ranchers	38

Table 1: Occupations With High Fatality Rates
Note. Data from National Census of Fatal Occupational Injuries in 2007 (USDL 08-1182), by Bureau of Labor Statistics, U.S. Department of Labor, 2008, Washington, DC: Author.
[a]per 100,000 workers

Opposition to Imposed Risks

Literature is abundant about people's resistance to being exposed to risks they believe are imposed on them. For some, the aversion to adopting the acceptable risk concept derives from their view that imposed risks are objectionable and are to be rebelled against. Conversely, they accept the significant risks of activities in which they choose to engage (e.g., skiing, bicycle riding, driving an automobile).

This idea needs exploration, which commences here with a statement that can withstand a test of good logic. As Stephans (2004) says, "The safety of an operation is determined long before the people, procedures, and plant and hardware come together at the worksite to perform a given task."

Start from the beginning in a process of creating a new facility and the credibility of Stephan's statement is validated. Consider, first, a site survey for ecological considerations, soil testing, then move into the facility's construction and fitting.

Thousands of safety-related decisions are made in the processes that result in an imposed level of risk. Usually, those decisions meet (or exceed) applicable safety-related codes and standards with respect to issues such as the contour of exterior grounds, sidewalks and parking lots; building foundations; facility layout and configuration; floor materials; roof supports; process selection and design; determination of the work methods; aisle spacing; traffic flow; hardware; equipment; tooling; materials to be used; energy choices and controls; lighting, heating and ventilation; fire protection; and environmental concerns.

Designers and engineers make decisions on these issues during the original design processes. Those decisions establish what the designers implicitly believe to be acceptable risk levels. Thus, the occupational and environmental risk levels have been largely imposed before a facility begins operation. Indeed, if those employed in such settings conclude that the imposed risks are not acceptable, communication systems should be in place to allow them to express their views and to have them resolved.

Minimum Risk as a Substitute for Acceptable Risk

Those who oppose use of the term acceptable risk often offer substitute terms. One frequent suggestion is to say that designers and operators should achieve *minimum* risk levels or *minimize* the risks. That sounds good, until one explores application of the terms.

Minimum means the least amount or the lowest amount. Minimization means to reduce something to the lowest possible amount or degree. Assume that the threshold limit value (TLV) for a chemical is 4 ppm. For $10 million, a system can be designed, built and installed that will operate at 2 ppm. For an additional $100 million, a 1 ppm exposure level can be achieved. Increase the investment to $200 million and the result is an exposure level of 0.1 ppm.

At 2 ppm, the exposure level is acceptable, but not minimum because a lower exposure level can be achieved. Requiring that systems be designed and operated to minimum risk levels, that risks be minimized, is impractical because the investments necessary to do so may be so high that the cost of the product required to recoup the investment and make a reasonable profit would not be competitive in the marketplace.

Designing to Standards as a Substitute for Acceptable Risk

Developing consensus standards often involves lively discussion, strong stances, much debate and many compromises. Some of these standards establish only minimum requirements. For example, the scope of ANSI/AIHA Z10-2005 states, "This standard defines minimum requirements for occupational health and safety management systems." Also, if a standard is obsolete, using it as a design base may result in designing to obsolescence and perhaps unacceptable risk levels.

Semiconductor Equipment and Materials International (2006) convincingly addresses the need to, sometimes, go beyond issued safety standards in the design process and to have decisions on acceptable risk levels be based on risk assessments.

> Compliance with design-based safety standards does not necessarily ensure adequate safety in complex or state-of-the-art systems. It often is necessary to perform hazard analyses to identify hazards that are specific with the system, and develop hazard control measures that adequately control the associated risk beyond those that are covered in existing design-based standards.

Designing to a particular standard may achieve an acceptable risk level, or it may not. In any case, the results of risk assessments and subsequent amelioration actions, if necessary, should be dominant in deciding whether acceptable risk levels have been reached.

Considerations in Defining Acceptable Risk

If the residual risk for a task or operation cannot be zero, for what risk level does one strive? Resources are always limited, and there is never enough money to address every hazard identified. As a result, SH&E professionals must give counsel so that the greatest good to society, employees, employers and product users is attained through applying available resources to obtain acceptable risk levels, practicably and economically.

Determining whether a risk is acceptable requires one to consider many variables. ISO/IEC Guide 51 (1999) speaks to the concept of designing and operating for risk levels as low as reasonably practicable.

> Tolerable risk [acceptable risk] is determined by the search for an optimal balance between the ideal of absolute safety and the demands to be met by a product, process or service, and factors such as benefit to the user, suitability for purpose, cost effectiveness and conventions of the society concerned.

Understanding cost effectiveness has become a more important element in risk acceptance decision making. That brings the discussion to ALARA and ALARP, commonly used acronyms in the risk assessment and applied risk reduction literature. ALARA stands for as low as reasonably achievable; ALARP stands for as low as reasonably practicable. Use of the ALARA concept as a guideline originated in the atomic energy field. According to Nuclear Regulatory Commission (2007):

> ALARA . . . means making every reasonable effort to maintain exposures to ionizing radiation as far below the dose limits as practical, consistent with the purpose for which

the licensed activity is undertaken, taking into account the state of technology, the economics of improvements in relation to benefits to the public health and safety, and other societal and socioeconomic considerations, and in relation to utilization of nuclear energy and licensed materials in the public interest (10 CFR 20.1003).

The implication that decision makers are to "[make] every reasonable effort to maintain exposures to ionizing radiation as far below the dose limits as practical" provides conceptual guidance in striving to achieve acceptable risk levels in all classes of operations.

ALARP seems to be an adaptation from ALARA. It has become the more frequently used term for operations outside the atomic arena and it appears more often in the literature. ALARP is that level of risk which can be further lowered only by an increment in resource expenditure that is disproportionate in relation to the resulting decrement of risk.

The concept embodied in these two terms applies to the design of products, facilities, equipment, work systems and methods, and environment controls. In the real world, benefits represented by the amount of risk reduction to be obtained and the costs to achieve those reductions are important factors. Trade-offs are frequent and necessary.

An appropriate goal in the decision-making process is to have the residual risk be ALARA. Paraphrasing the terms contained in the definition of ALARA helps explain the process:

1) Reasonable efforts are to be made to identify, evaluate, and eliminate or control hazards so that the risks deriving from those hazards are acceptable.

2) In the design and redesign processes for physical systems and for the work methods, risk levels for injuries and illnesses, and property and environmental damage, are to be as far below what would be achieved by applying current standards and guidelines as is economically practicable.

3) For items 1 and 2, decision makers are to consider purpose of the undertaking; state of the technology; costs of improvements in relation to benefits to be obtained; and whether the expenditures for risk reduction in a given situation could be applied elsewhere with greater benefit.

Since resources are always limited, spending an inordinate amount of money to reduce the risk only slightly through costly engineering and redesign is inappropriate, particularly if that money could be better spent elsewhere. This premise can be demonstrated through an example that uses a risk assessment matrix as a part of the decision making.

Risk Assessment Matrix

A risk assessment matrix that assigns numbers to risk levels demonstrates the application of the ALARP principle. One must understand that the numbers in the

	Occurrence probability and values				
Severity level and values	Very low (1)	Low (2)	Moderate (3)	High (4)	Very high (5)
Very high (5)	5	10	15	20	25
High (4)	4	8	12	16	20
Moderate (3)	3	6	9	12	15
Low (2)	2	4	6	8	10
Very low (1)	1	2	3	4	5

Incident or exposure probability descriptions
Very low: Improbable, very unlikely
Low: Remote, may occur, but not likely
Moderate: Occasional, likely to occur sometime
High: Probable, likely to occur several times
Very high: Frequent, likely to occur repeatedly

Incident or exposure severity descriptions
Very low: Inconsequential with respect to: injuries or illnesses, system loss or down time, or environmental chemical release
Low: Negligible: first aid or minor medical treatment only, non-serious equipment or facility damage, chemical release requiring routine cleanup without reporting
Moderate: Marginal: medical treatment or restricted work, minor subsystem loss or damage, chemical release triggering external reporting requirements
High: Critical: disabling injury or illness, major property damage and business down time, chemical release with temporary environmental or public health impact
Very high: Catastrophic: one or more fatalities, total system loss, chemical release with lasting environmental or public health impact

Risk scoring and categories
Combining probability values with severity descriptions yields a risk score. That score can be categorized as follows.

Risk score
Under 4   Category 1: Remedial action discretionary
4 to 8   Category 2: Remedial action to be taken at appropriate time
9 to 14   Category 3: Remedial action to be given high priority
15 or greater   Category 4: Operation not permissible. Immediate action necessary

Table 2: Risk Assessment Matrix

matrix presented (Table 2) are qualitative, not quantitative. They are relational and have meaning as they interact with each other. Many other risk assessment matrixes could be used as well. An SH&E professional may want to use other probability and severity descriptions and risk scoring categories. Combining the severity and occurrence probability values yields a risk score in the matrix. Table 2 also includes information on categorizing the risks and action levels based on urgency.

The following example illustrates how a team used the matrix and applied the ALARP concept to make a decision about acceptable risk.

1) A chemical operation was built 15 years ago. While engineering modifications have been made in the system over the years, management knows that its operations are no longer state of the art.

2) A risk assessment team is convened to consider the chemically related risks in a particular process in the overall system.

3) In the deliberations, the group refers to its established hierarchy of controls:

a) Eliminate or reduce risks in the design and redesign processes.

b) Reduce risks by substituting less hazardous methods or materials.

c) Incorporate safety devices.

d) Provide warning systems.

e) Apply administrative controls (e.g., work methods, training, work scheduling).

f) Provide PPE.

4) The group first considers the possibility of redesigning and replacing the process. Substitution of materials or methods is considered, but the group determines that such opportunities have already been addressed. Safety devices and warning systems are considered state of the art, and maintenance is considered superior.

5) Occurrence probability for a chemically related illness is judged to be moderate (3) and the severity level is moderate (3). Thus, the risk score is 9, which is in Category 3 and remedial action is to be given high priority.

6) The team recognizes that to reduce the risk further, appropriate training must be delivered and repeated, and standard operating procedures and the use of PPE must be rigidly enforced.

7) Management agrees to fund the necessary administrative improvements.

8) Assuming that these improvements are made, the risk assessment group decides that the probability of occurrence of an illness from a chemical exposure would be low (2) and that the severity of harm expected would be low (2). Thus, the risk score is 4, in Category 1.

9) Reengineering and replacing the process would reduce the probability level to very low (1) and the severity level to very low (1), thereby achieving a risk score of 1,

also is in Category 1. The estimated cost of redesigning and replacing the process, $1.5 million, was considered disproportionate with respect to the amount of risk reduction to be obtained.

10) The risk assessment group tells management that it would prefer having the money spent on a wellness center.

The ALARP Principle

ALARP promotes a management review, the intent of which is to achieve acceptable risk levels. Practical, economic risk trade-offs are frequent and necessary in the benefit/cost deliberations that occur when determining whether the costs to reduce risks further can be justified "by the resulting decrement in risk."

Several depictions of the ALARP concept begin with an inverted triangle (Figure 1) because it indicates that risk is greater at the top and much less at the bottom. Figure 1 shows the concept combined with elements in the risk assessment matrix.

Defining Acceptable Risk

This author's definition of acceptable risk is included in ASSE TR-Z790.001-2009. Risk acceptance is a function of many factors and varies considerably across industries (e.g., mining vs. medical devices vs. farming). Even at locations of a single global company, acceptable risk levels can vary. Company culture and the culture of the country in which a facility is located influence risk acceptability, according to colleagues working in global companies. Training, experience and resources also can influence acceptable risk levels. Risk acceptability is also time dependent, in that what is acceptable today may not be acceptable tomorrow, next year or the next decade.

A sound, workable definition of acceptable risk must encompass hazards, risks, probability, severity and economic considerations. This author believes that the definition of acceptable risk included in this article represents the development of and practical use of the term over the past several years.

Social Responsibility: An Emerging Opportunity

Formal consideration of social responsibility by senior executives is a fairly recent development. What is social responsibility? An Internet search will reveal a large number of definitions. This article focuses on two.

1) The World Business Council for Sustainable Development (2000) defines corporate social responsibility as "the continuing commitment by business to behave ethi-

Chapter 8: Acceptable Risk

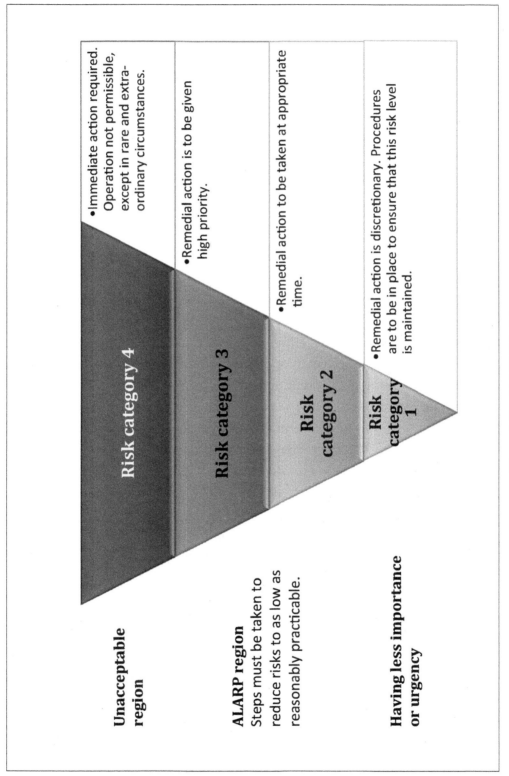

Figure 1: The ALARP Principle

cally and contribute to economic development while improving the quality of life of the workforce and their families as well as of the local community and society at large."

2) Gap Inc. states, "[S]ocial responsibility is fundamental to show how we do business. It means everything from ensuring that workers are treated fairly to addressing our environmental impact."

It is logical to suggest that if a company initiates a social responsibility endeavor which is to include the well-being of workers, the environment and the community at large, knowledge and application of acceptable risk principles would inform its decision making. The result would be efficient allocation of resources, fewer injuries and illnesses and property damage incidents, and serving the community well. That seems to present opportunities for SH&E professionals.

The State of the Art in Risk Assessment

SH&E professionals must understand that risk assessment is as much an art as science and that subjective judgments—educated, to be sure—are made on incident or exposure probability and the severity of outcome to arrive at a risk category. Also, one must recognize that economically applicable risk assessment methodologies have not been developed to resolve all risk situations.

For example, when asked, "How would you assess the cumulative risk in an operation in which there was an unacceptable noise level and toluene was used in the process?" one would hope that resource material such as EPA's (2003) *Framework for Cumulative Risk Assessment* would provide an answer. It does not. The agency is cautionary about cumulative risk assessment methods.

> It should be acknowledged by all practitioners of cumulative risk assessment that in the current state of the science there will be limitations in methods and data available (p. 31).
>
> Finding a common metric for dissimilar risks is not an analytical process, because some judgments should be made as to how to link two or more separate scales of risks. These judgments often involve subjective values, and because of this, it is a deliberative process (p. 55).
>
> Calculating individual stressor risks and then combining them largely presents the same challenges as combination toxicology but also adds some statistical stumbling blocks (p. 66).

Where multiple, diverse hazards exist, the practical approach is to treat each hazard independently, with the intent of achieving acceptable risk levels for all. In the noise and toluene example, the hazards are indeed independent. Complex situations, or when evaluating competing solutions to complex systems, may require the assistance of specialists with knowledge of more sophisticated risk assessment methodologies such as hazard and operability analysis or fault tree analysis. For most applications, however, the author does not recommend that diverse risks be summed through what could be a questionable methodology.

Conclusion

Risk acceptance is the deliberate decision to assume a risk that is low enough with respect to the probability of a hazard-related incident or exposure occurring and the severity of harm or damage that may result, and which is considered tolerable in a given situation. Management's decision to accept a risk should be deliberate and the criteria for the decision should be documented. In an ideal world, all personnel who are impacted should be involved in or be informed of risk acceptance decisions.

Use of the term acceptable risk has arrived. It is becoming a norm. In organizations with advanced safety management systems, the idea of achieving practicable and acceptable risk levels throughout all operations is a cultural value. It is suggested that SH&E professionals adopt the concept of attaining acceptable risk levels as a goal to be embedded in every risk elimination or reduction action proposed. To achieve that goal, SH&E professionals must educate others on the benefits of applying the concept.

SH&E professionals also must be able to work through the greatly differing views people can have about risk levels, incident and exposure probabilities, and severity. Workers may have differing views about risk and they should be considered for their value. With respect to environmental risks, community views must be considered as well.

In arriving at acceptable risk levels where the hazard/risk scenarios are complex, it is best to gather a team of experienced personnel for their contributions and for their buy-in to the conclusions. ■

References
ANSI/AIHA. (2005). American national standard for occupational health and safety management systems (ANSI/AIHA Z10-2005). Fairfax, VA: Author.
ANSI/ASSE. (2003/2009). American national standard for control of hazardous energy: Lockout/tagout and alternative methods (ANSI/ASSE Z244.1-2003(R2009). Des Plaines, IL: Author.
ANSI/Association for Manufacturing Technology. (2000). Risk assessment and risk reduction: A guide to estimate, evaluate and reduce risks associated with machine tools (ANSI B11.TR3-2000). McLean, VA: Author.
ANSI/Packaging Machinery Manufacturers Institute (PMMI). (2006) American national standard for safety requirements for packaging machinery and packaging-related converting machinery (ANSI/PMMI B155.1-2006.) Arlington, VA: Author.

ASSE. (2009). Prevention through design: Guidelines for addressing occupational risks in design and redesign processes (ASSE TR-Z790.001-2009). Des Plaines, IL: Author.

British Standards Institution (BSI). (2007). Occupational health and safety management systems: Requirements (BS OHSAS 18001:2007). London: Author.

Bureau of Labor Statistics. (2008, Aug. 20). *National census of fatal occupational injuries in 2007* (USDL 08-1182). Washington, DC: U.S. Department of Labor, Author.

DOT. (2005). Risk management definitions. Washington, DC: Author, Pipeline and Hazardous Materials Safety Administration. Retrieved March 19, 2010, from http://www.phmsa.dot.gov/hazmat/risk/definitions.

EPA. (2003, May). Framework for cumulative risk assessment (EPA/630/P-02/001F). Washington, DC: Author. Retrieved March 19, 2010, from http://cfpub.epa.gov/ncea/cfm/recordisplay.cfm?deid=54944.

Fewtrell, L. & Bartram, J. (Eds.). (2001). *Water quality: Guidelines, standards and health*. London: IWA Publishing for World Health Organization.

Gap Inc. Social responsibility. San Francisco: Author. Retrieved March 19, 2010, from http://www.gapinc.com/public/SocialResponsibility/socialres.shtml.

Health and Safety Executive. ALARP "at a glance." London: Author. Retrieved March 19, 2010, from http://www.hse.gov.uk/risk/theory/alarpglance.htm.

Industrial Union Department, AFL-CIO v. American Petroleum Institute U.S. at 642.

Institute for Research for Safety and Security at Work & The Commission for Safety and Security at Work in Quebec. (2009). Machine safety: Prevention of mechanical hazards. Quebec, Canada: Author. Retrieved March 19, 2010, from http://www.irsst.qc.ca/files/documents/PubIRSST/RG-597-Pref-TCont-Intr.pdf.

International Electrotechnical Commission (IEC). (2007). International standard for environmentally conscious design of medical equipment (IEC 60601-1-9). Geneva: Author.

International Organization for Standardization (ISO)/IEC. (1999). Safety aspects: Guidelines for their inclusion in standards [ISO/IEC Guide 51:1999(E)]. Geneva: Author.

Lowrance, W.F. (1976). *Of acceptable risk: Science and the determination of safety*. Los Altos, CA: William Kaufman Inc.

Manuele, F.A. (2008). *Advanced safety management: Focusing on Z10 and serious injury prevention*. Hoboken, NJ: John Wiley & Sons.

Manuele, F.A. (2003). *On the practice of safety* (3rd ed.). Hoboken, NJ: John Wiley & Sons.

Manuele, F.A. & Main, B. (2002, Jan.). On acceptable risk. *Occupational Hazards*. Retrieved March 19, 2010, from http://ehstoday.com/news/ehs_imp_35066.

Nuclear Regulatory Commission. (2007). 20.1003 Definitions: ALARA. Washington, DC: Author. Retrieved March 19, 2010, from http://www.nrc.gov/reading-rm/doc-collections/cfr/part020/part020-1003.html.

OSHA. (2003). Voluntary Protection Programs: Policies and procedures manual (archived). Washington, DC: U.S. Department of Labor, Author. Retrieved March 19, 2010, from http://www.osha.gov/pls/oshaweb/owadisp.show_document?p_table=DIRECTIVES&p_id=2976.

Presidential/Congressional Commission on Risk Assessment and Risk Management. (1997). *Framework for environmental health risk management.* Washington, DC: Author.

Semiconductor Equipment and Materials International (SEMI). (2006). Environmental, health and safety guideline for semiconductor manufacturing equipment (SEMI S2-0706). San Jose, CA: Author.

SEMI. (1996). Safety guideline for risk assessment (SEMI S10-1996). Mountain View, CA: Author.

Standards Association of Australia. (2004). Risk management standard (AS/NZS 4360: 2004). Strathfield, NSW, Australia: Author.

Stephans, R.A. (2004). *System safety for the 21st century.* Hoboken, NJ: John Wiley & Sons.

United Nations/International Strategy for Disaster Reduction (UN/ISDR). (2009). UNISDR terminology on disaster risk reduction. New York: Author. Retrieved March 19, 2010, from http://www.unisdr.org/eng/library/lib-terminology-eng.htm.

World Business Council for Sustainable Development. (2000). Corporate social responsibility: Making good business sense. Geneva: Author. Retrieved March, 19, 2010, from http://www.wbcsd.org/DocRoot/IunSPdIKvmYH5HjbN4XC/csr2000.pdf.

9 Accident Costs
Rethinking Ratios of Indirect to Direct Costs

ORIGINALLY PUBLISHED JANUARY 2011

Presentations to management on the costs of worker injuries and illnesses can be attention-getting and convincing, provided the data are plausible and can be supported with suitable references. Unfortunately, little research and hard data exist to support the frequently used ratios of indirect to direct costs that appear in safety-related literature.

Furthermore, as in the sources cited by this article, the elements included in direct and indirect cost categories may differ (Heinrich, 1931; Grimaldi & Simonds, 1989; Leigh, Markowitz, Fahs, et al., 1997). And, the ratios in those sources are invalid because the direct costs of accidents have increased in recent years at a pace far greater than indirect costs.

This article discusses the author's review of select data pertaining to indirect and direct accident costs. Computations are made in order to update a ratio reported in a plausible research study in order to approximate the current ratio of indirect to direct costs. In addition, the author discusses the inappropriateness of the "additional sales needed" argument to cover total indirect and direct accident costs.

Unsupported Statements About Ratios

Enter "indirect and direct costs of accidents" into an Internet search engine, and the search will return a wide variety of documents that include ratios of those costs.

Some pertain to the costs that an employer would bear. Others are more broad and pertain to the societal burden of such costs. Select examples relating to employer costs are highlighted.

- *The Business Results Through Health and Safety Guidebook*, from Canadian Manufacturers and Exporters (Ontario Division) and Workplace Safety and Insurance Board (2001) "demonstrates the business case for workplace health and safety and reflects the experience of Ontario businesses." The publication states, "The average workplace lost-time injury in Ontario costs over $59,000. The average lost-time workers' compensation claim cost is over $11,771." (Note: Round $11,771 to $11,800, and one finds that a 4-to-1 multiplier was used to get to $59,000. The guidebook recommends a 4:1 ratio within a cost computation system provided for employers to use.)

- The Spring 2006 issue of ASSE's *Journal of SH&E Research* contains the article, "A Survey of the Safety Roles and Costs of Injuries in the Roofing Contracting Industry" Choi (2006). The author writes, "Traditionally indirect costs are measured as being four times the direct costs (Heinrich, 1941), but the indirect costs of injuries may range from two to 20 times the direct costs."

- U.S. Fish and Wildlife Service, Division of Safety and Health, offers this: "For every dollar spent on direct costs, $4 to $10 are spent on indirect costs."

- Western National Insurance says, "Most experts estimate that the indirect costs are 3 to 10 times the direct costs of an accident."

- North Carolina Industrial Commission (2007) states, "Many seasoned experts estimate that the indirect costs of an accident are three to 10 times the direct costs."

- International Labor Organization's Introduction to Occupational Safety and Health training module states, "It has been estimated that the indirect costs of an accident or illness can be four to 10 times greater than the direct costs, or even more."

- International Safety Equipment Association (ISEA, 2002) says, "Reliable estimates place them (indirect costs) at up to 30 times the direct costs."

- OSHA (2007) indicates that "studies show that the ratio of indirect costs to direct costs varies widely, from a high of 20:1 to a low of 1:1."

Some of these references say the ratios relate to the work of seasoned experts, reliable estimates or studies made, but none of the experts or studies are cited. The ratios cited in these examples range from 1:1 to 30:1.

Differences in Cost Categories

Many combinations of terms about the costs to employers of employee accidents appear in the literature, including direct and indirect, insured and uninsured, ledger and nonledger, and tangible and intangible. (Ledger and tangible costs are those for which data are created in the normal business process that can be entered into a financial ledger, such as medical and indemnity costs paid. Nonledger and intangible costs are those that occur but for which no specific data are determined, such as the value of time spent by supervisors and others who give attention to the accident, time spent providing first aid and investigation time.)

Also, the direct and indirect cost categories differ considerably among the various sources. What is an indirect or hidden or uninsured cost in one list may be excluded in another. For example, some lists include lost productivity or loss of profits as an indirect cost. Chapter 6, "Cost Analysis," in Grimaldi and Simonds (1989) *Safety Management*, 5th ed., contains a 6-page discussion on the invalid and restricted items that should not be included when computing uninsured costs. For example:

> Loss of profit of idle machines or workers is not generally a valid cost. When workers or machines are made idle, one of two things occurs. Either the production is eventually made up, or it is not. If it is made up, in the sense that over a long period of time no less goods are produced and sold than would have been had the "accident" not occurred, there is no loss of profit, apart from the increase in production costs.

For this article, the author focused on accident costs assumed by employers, the direct costs of which are the legally required indemnity payments and the medical costs paid, with all other related costs being the indirect costs. Other studies have addressed the costs of injuries and illnesses to society; however, because of the differences in cost allocation methods, those studies are of little value in determining employer costs.

Heinrich's Indirect and Direct Cost Ratios

Heinrich's (1931; 1959) presentation of the 4:1 indirect to direct cost ratio of injuries and illnesses was a historical first. It is often referenced in the literature. The following comments are derived from the first edition of the book (published in 1931), giving the results of his 1926 study.

Heinrich wrote that, according to his research and analysis, an employer's cost of the so-called "incidental" costs of worker injuries was four times as great as compensation and medical payments. This 4:1 ratio of indirect to direct costs also appeared in the three later editions of his book. Many statements in safety-related literature repeat Heinrich's 4-to-1 ratio. Although his studies were made in 1926, his ratio has had staying power.

Heinrich's direct costs of injuries and illnesses are "compensation and liability claims, medical and hospital cost, insurance premiums, and cost of lost time except when actually paid by the employer without reimbursement." His list of hidden accident cost factors includes 11 subjects, some of which have subparts. Only the major captions appear here; some have been combined:

- time of the injured employee, the other employees who stop work or who are upset;
- time of foremen, supervisors or other executives who give attention to the injury;
- time spent by first-aid attendants and hospital department staff when not paid for by the insurance carrier;
- damage to any machines, equipment and other property, and interference with the site's production;
- costs to the employer of welfare and benefit systems, and continuing the full wages of the employee after returning to work that are not fully recovered;
- loss of profit on the injured person's production, on idle machines;
- overhead (e.g., lights, heat) when the employee is away or not fully productive.

In the third and fourth editions of his book, Heinrich included data on nine cases to support his ratio (fewer in the earlier editions). This author has found no other data that authenticate his research. Also, Heinrich said, "The examples of hidden costs given in this chapter include no fatalities, major dismemberments or major permanent injuries, nor do they feature spectacular costs that result from trivial injuries." Thus, the resulting ratios he presents are limited to the less serious injuries.

Keep in mind that Heinrich's analysis was conducted in 1926. How valid could any ratios be in 2011 after 85 years of immense change in industry and business, work practices and compensation systems, and which do not consider the advances made in the practice of medicine and increases of indirect and direct costs?

Bird on Accident Costs

Bird (1974) presents the so-called iceberg theory of incident costs. In an exhibit that suggests an iceberg and is captioned "The Real Costs of Accidents Can Be Measured and Controlled," Bird illustrates his ratios of insured and uninsured costs. Bird and Germain (1985) use these same ratios but add to the descriptions of what costs are included in each category. In Bird's data, the insured costs—medical and compensation costs—are the same as direct costs in some other presentations.

The 1985 version is the base of an adaptation of the iceberg (Figure 1). As a minimum, Bird's data imply that employers absorb uninsured costs at a ratio of $6 (5+1) to $1 of insured costs. At a maximum, the ratio is $53 (50+3) to $1. This author has located no research or hard data to support such ratios. Depictions of Bird's iceberg have appeared in many texts and articles, and his cost ratios are frequently repeated.

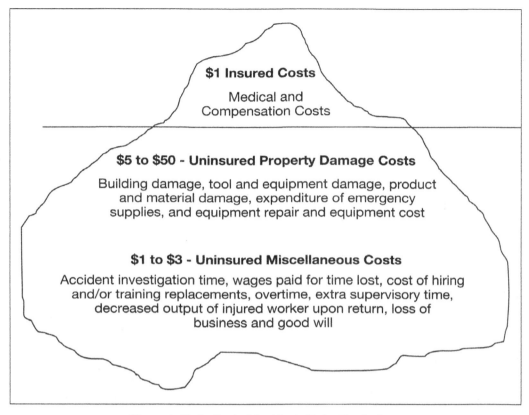

Figure 1: Bird's Cost of Accidents Data: The Iceberg
Note. Adapted from *Practical Loss Control Leadership* (Revised ed.), by F.E. Bird Jr. and G.L. Germain, 1985, Loganville, GA: Det Norske Veritas.

Bird's assertion that uninsured property damage costs are 5 to 50 times the insured costs are difficult to support universally, but such high ratios could exist in an organization that experiences an incident which causes significant property damage. Over a period of years, the impact of property damage costs from hazards-related incidents on indirect costs will relate directly to the investment per employee in the physical plant. Consider these two extremes.

•At a public utility, the cost of the investment in the physical plant is $1.2 million per employee. A turbine rotor was being hoisted to be returned to its bearings when a wire cable sling broke. For an injured worker, the total workers' compensation cost was $2,400; the cost of the damage to the rotor was $6.5 million.

•In a shop where employees sew shirts, the investment per worker is about $6,000. No high-cost equipment exists to be damaged, and an indirect cost ratio many times the direct costs for businesses of this nature is not plausible.

A Construction Industry Study

In August 1981, under contract to The Business Roundtable, Stanford University's Department of Civil Engineering issued Technical Report No. 260, "Improving Construction Safety Performance: The User's Role." This research is the latest study located by the author. It was conducted to provide guidance on reducing accident frequency and severity in the construction industry, and the attendant indirect and direct injury and illness costs.

In 1982, The Business Roundtable issued a condensed version of the report, "Improving Construction Safety Performance." This version does not include the research methodology; the questionnaire developed and sent to contractors; the instructions given to contractors on how to identify direct and indirect costs; or the analytical data.

The Stanford study is important work since it establishes significant data points on indirect and direct injury and illness costs. Keep in mind that the dollar amounts in the report pertain to 1980-81 when the research was conducted. Interesting discussion could take place about the method, the definitions of terms, the content of the data collection forms and the data collection system. Nevertheless, the work is deemed significant because:

1) A study was actually conducted.

2) The report confirms the lack of data to support other published ratios.

3) Research determined that as injury severity increased, the ratio of indirect to direct costs decreased.

4) The indirect cost ratio is affected by many variables and establishing a ratio of indirect to direct costs that is universally applicable to all entities at all times is not feasible, particularly because of the great variations that could occur in the cost of property damage which could result from an accident.

5) Computations indicate that, for this study, the indirect to direct cost ratio is 1.6 to 1.

The following comments expand on the foregoing observations.

1) A study was actually conducted, following accepted research practices, and the results were documented in detail. A questionnaire was developed to gather information, an instruction guide was prepared and sent to contractors, and data were gathered and analyzed. Thus, research was performed, and a report covering the data gathering, and analysis system and the conclusions drawn is available for review.

2) The report confirms the lack of data to support other published ratios. It states:

> For this aspect of the research [Hidden Cost of Accidents], much effort was expended trying to determine the source of the ratios that have appeared in various reports. No hard data [were] uncovered to substantiate any of the widely publicized ratios that had come to the attention of the researchers (Stanford, 1981).

Chapter 9: Accident Costs

To emphasize: No hard data were found in support of other statistics on the indirect to direct cost ratio of injuries published prior to 1981.

3) The research determined that as injury severity increased, as represented by higher workers' compensation costs, generally, the ratio of indirect costs to direct costs decreased (Table 1).

Range of benefits paid ($)	No. of cases	Average benefits paid ($)	Average hidden cost ($)	Average ratio Hidden cost: Benefits paid
No lost time				
0 to 199	13	125	530	4.2
200 to 399	7	250	1,275	5.1
400+	4	940	4,740	9.2
Lost time				
0 to 2,999	9	869	3,600	4.1
3,000 to 4,999	8	3,694	6,100	1.6
5,000 to 9,999	4	6,602	7,900	1.2
10,000+	4	17,137	19,640	1.1

Table 1: Accident Costs: Stanford Report
Note. Analysis of accident costs by size of accident: Stanford research.
Adapted from "Improving Construction Safety Performance: The User's Role" (Technical Report No. 260), by Stanford University, 1981, Palo Alto, CA: Author, Department of Civil Engineering.

4) Analysis of the compiled data showed that the indirect cost ratio is affected by many variables and that attempting to establish a ratio of indirect to direct costs that could be universally applied in all operations at all times is not feasible. "Still, if the accident data are separated into two general groups, large and small claims, it can be seen that smaller accidents have a larger [indirect to direct cost] multiplier."

5) The total of hidden costs for all claims was divided by the total cost for benefits paid, resulting in a 1.6:1 ratio of indirect to direct costs. But the report cautions that the 1.6:1 ratio may be low because certain indirect cost data were omitted, such as "OSHA fines and hearings and third-party liability and legal actions, and the effect of accident costs on future workers compensation premiums."

Table 1 is from the Stanford report. It is a summation of data obtained from contractors on each injury for which data are recorded in the analytical forms. In this study, the hidden cost multiplier was defined "to account for all costs other than the direct compensation of the victim(s)." That is, other than for the indemnities paid to injured workers and the attendant medical costs. "Hidden costs" are: insurance company claims handling and administration costs; other wages; efficiency loss; rehabilitation; supervisor costs; transportation; overtime costs; break-in replacement; materials, equipment, clean up and related costs; clerical costs; and other.

The data in Table 1 support the premise that as injury severity increases, the indirect cost ratio decreases. The Business Roundable report states, "The ratio [of the indirect to direct cost] varies greatly with the magnitude of the accident; however, it is not necessarily linked to the severity of the injury." The premise that as injury severity increases the ratio of indirect to direct costs decreases will withstand a logic test for various classes of injury costs.

The highest value of benefits paid for an injury in the Stanford study was $24,900. One can speculate that decreases of the ratio would continue incrementally as claim costs increased for cases valued at $25,000 to $50,000, from $50,000 to $100,000, and so on.

A Cost Distribution Study

The data in Table 2 were produced, confidentially, for this author by a third-party administrator for the total cost of workers' compensation claims, paid and reserves. As will be noted, cases valued at $25,000 or more represent 6% of the total number of cases and 72% of the total costs. If data were included in the Stanford study on a significant number of cases valued at $25,000 to $50,000 and more than $50,000, and a continuing decrease in the indirect to direct cost ratio occurred as the dollar value of the cases increased, the composite indirect to direct cost ratio—1.6:1—reported in the Stanford study would be much lower. However, that is speculation for which hard data do not exist.

Given the types of indirect cost categories listed in the cited studies, one can conclude that a large share of the indirect costs is expended in the 3 or 4 weeks immediately following the injury. This is an important point. For the very severe injury, say valued at $500,000 for workers' compensation cost and for which long-term disability will result, it is not reasonable to predict that the indirect cost will be at the 1.1:1 ratio for that claim value as shown in the Stanford study. That indirect cost computes at $550,000.

Claim value	Percentage of no. of claims	Percentage of total costs
$0 to $10,000	89%	14%
$10,000 to $25,000	5%	14%
$25,000 to $50,000	3%	20%
More than $50,000	3%	52%
$10,000 or more	11%	86%
$25,000 or more	6%	72%

Table 2: Workers' Compensation Claims
Note. Computer run of 280,000 workers' compensation claims for 2003.

Updating the Stanford Indirect and Direct Cost Ratios

The relationship between indirect to direct costs has changed significantly since 1980. The following exercise was devised to estimate what the 1.6-to-1 multiplier from the Stanford study might be as of 2008. Understand that these computations are illustrative only. They were made to show, generally, how significantly the ratios may have changed. Nevertheless, they are helpful and instructive. Should statisticians delve more precisely into the trending of indirect and direct costs, their computations could achieve a higher accuracy level.

BLS's online inflation calculator uses the average Consumer Price Index for a year as a basis for data. Table 3 presents the results of these computations.

Table 4 presents data from the "2009 State of the Line" report (Mealy, 2009) from National Council on Compensation Insurance (NCCI). Indirect costs are primarily the costs of time spent by persons other than the injured person, and property damage. Assume that indirect costs increased from 1980 through 2008 at the same rate as inflation: 2.61. For lost-time cases in the period from 1995 through 2008, the combined indemnity and medical claims costs increased about 3.27 times the inflation rate (134/41).

Amount spent	Year	Had the same purchasing power as	In the year	Multiplier
$1.00	1980	$2.61	2008	2.61
$1.00	1980	$1.80	2004	1.80
$1.00	1995	$1.41	2008	1.41

Table 3: Inflation Computations
Note. Inflation computations relating purchasing power in 1 year to a subsequent year.

Average indemnity cost per lost-time claim		
1995	$10,500	
2008	$22,500	114.3% increase
Average medical cost per lost-time claim		
1995	$9,800	
2008	$25,200	157.1% increase
Total: Indemnity and medical cost per lost-time claim		
1995	$10,500 + $9,800 = $20,300	
2008	$22,500 + $25,000 = $47,500	134% increase

Table 4: Progression With Respect to Workers' Compensation Claims Costs
Note. Data from "2009 State of the Line," by D.C. Mealy, 2009, Boca Raton, FL: National Council on Compensation Insurance.

Assume that the 134% increase also applies to no-lost-time cases. Then, assume that workers' compensation claims costs (the direct costs) increased overall at about the same rate as did inflation from 1980 through 1994: $1 to $1.80, and from $1 to $1.34 for 1995 through 2008 ($1.80 + $1.34 = $3.14)—more than three times the inflation rate.

The new relationship of indirect to direct costs, reflecting increases from 1980 to 2008:
- Direct cost: $3.14
- Indirect cost: $2.61
- 2.61/3.14 = 0.81 (adjusted to 0.8)
- Ratio: 0.8 to 1 (an approximation)

This ratio reflects the significant differences in indirect and direct cost increases. Computations show approximately a 49% reduction of the 1.6:1 average ratio shown in the Stanford report. As workers' compensation healthcare costs continue to increase at a rate considerably higher than inflation, the value of the ratios in the Stanford report will diminish further.

A NIOSH Research Project

A team of researchers under contract with NIOSH conducted an extensive investigation to produce a 1997 report, "Costs of Occupational Injuries and Illnesses in 1992" (Leigh, Markowitz, Fahs, et al., 1997) Its title might suggest that the report would be valuable as a reference in determining indirect and direct injury and illness costs borne by an employer.

However, this report was produced for a different purpose: to establish the costs to society as a whole. Therefore, the data categories differ, which minimizes the report's usefulness as a resource to determine employer costs.

As would be typical in such a study (so the author was advised by an economist), these researchers took an important departure from a basic premise in the Heinrich, Bird and Stanford data. In each of those studies, the workers' compensation costs, namely, indemnity paid to the worker and the attendant medical costs, are the direct costs. Not so in this 1997 report. Many of the costs included as direct costs in this report do not appear as direct costs in other studies. According to the NIOSH report:

> Direct costs represent actual dollars spent or anticipated to be spent on providing medical care to an injured or ill person as well as property damage, police and fire services, administrative costs for delivering indemnity benefits and direct costs to innocent third parties. Medical costs include doctors' and nurses' services, hospital charges, drug costs, rehabilitation services, ambulance fees, payments for medical equipment and supplies.
>
> Indemnity benefit costs do not include the benefits themselves; these are implicitly accounted for in the indirect costs (lost wages). Indemnity benefit costs include the ad-

ministration costs associated with providing workers' compensation indemnity or Social Security disability payments to injured or sick workers and their families. Property damage includes costs of damage to vehicles, machines, buildings and so on, directly associated with the injuries and illnesses.

The largest indirect costs include the injured or sick worker's lost earnings, fringe benefits and home production. Other indirect costs include employer's costs associated with retraining and restaffing, coworker costs of lost productivity, time delays and indirect costs to innocent bystanders (Leigh, et al., 1997).

Because of the allocation of costs into direct and indirect categories, the ratios developed are not comparable with any other system discussed in this article. Much of the NIOSH report is devoted to occupational illnesses and to speculations about their indirect and direct costs. Nevertheless, the report concludes:

> Identifying the costs of occupational injury and disease in the U.S. has been an elusive goal. Our study represents the first attempt to estimate the national cost of occupational injury and illness using national data. But, even with our study, this goal remains unattained (Leigh, et al., 1997).

The case is made that because the data required for a proper study do not exist, many speculative assumptions must be made to produce speculative conclusions.

OSHA Adopted the Stanford Report Ratios

OSHA gives credibility to the ratios that arose out of the Stanford research by publishing and promoting use of the data in The Business Roundtable version of the Stanford report, with some revision. The data in Table 5 appear in OSHA's Safety and Health Management Systems eTool (2007).

Direct cost of claim	$0-2,999	$3,000-4,999	$5,000-9,999	$10,000+
Ratio of indirect to direct cost	4.5	1.6	1.2	1.1

Table 5: OSHA's Ratio of Indirect to Direct Costs
Note. Data from "Safety and Health Management Systems eTool: Costs of Accidents," by OSHA, 2007, Washington, DC: Author.

OSHA acknowledges the source of its data as "Business Roundtable, Improving Construction Safety Performance: Report A-3, January, 1982." The indirect cost ratio for the $0-2,999 "Range of Benefits Paid" in the Business Roundtable report is 4.1:1, rather than 4.5 as shown in Table 5. However, note that the OSHA exhibit shows no data for "no-lost-time" claims, for which the composite ratio would be higher than 4.1. This indicates that some combinations were made. The eTool says:

•Workers' compensation claims which cover medical costs and indemnity payments for an injured or ill worker are the direct costs.
•All other related costs are the indirect costs.
•To help assess the impact of occupational injuries and illnesses on your profitability, try out OSHA's "$afety Pays" program. It uses a company's profit margin, the average cost of an injury or illness, and an indirect cost multiplier to project the number of sales you would need to cover these costs (OSHA, 2007).

An OSHA bulletin on its $afety Pays program, "Do You Know How Much Accidents Are Really Costing Your Business?" includes the ratios in Table 5 as well as a depiction of Bird's iceberg. Another bulletin (OSHA, 2009a) says that the program will:
•offer choices from a set of lost workday injuries and illnesses;
•prompt users for information to do the analysis;
•allow users to input the actual loss figures or workers' compensation costs;
•generate a report of the costs and the sales needed to cover those costs.

The injury and illness types and average costs in the OSHA set were provided by NCCI for 2004 lost-time cases only. The database is large and the costs for the injury and illness categories have validity for that year. The database has 53 injury and illness categories. The average cost for each category is provided along with the appropriate indirect cost multiplier: For 52 of the categories, the indirect cost ratio is the same: 1.1 to 1. For the 53rd category, the ratio is 1.2 to 1.

The 1.1-to-1 ratio is shown in the Stanford report for claims valued at $10,000 or more. For the 1.2-to-1 ratio, the claims value is $5,000 to $9,999. Of the 53 injury and illness categories, 32 are valued above $25,000; they range from $25,004 to $115,961. Applying a 1.1-to-1 indirect to direct cost ratio to such higher valued claims is questionable.

The calculations performed to relate costs in 1980-81 to current costs resulted in a 49% reduction in the 1.6-to-1 ratio, which appears in the Stanford report as the composite ratio for all injury levels. Reducing the 1.1-to-1 ratio used in OSHA's $afety Pays program by 49% produces a revised ratio of 0.56 (rounded to 0.6) to 1. Data on indirect costs produced using this program is misleading.

Computing the "Additional Sales" Necessary to Cover Injury Costs

Bird (1974) proposes that reports to management should include data to show the "additional sales" required to pay for accident costs. Needing additional sales to cover indirect and direct costs is the subject being questioned. The following discussion shows that the premise that additional sales are necessary to cover accident costs cannot be supported.

An executive with whom this subject was reviewed and whose background is in finance suggested that relating total accident costs to dollars of profit could be conceptually supported and such a comparison might have significance. For example, a report prepared by a safety professional for management indicating that injury costs are equal to 100% of profits could get attention.

Bird (1974) offers a chart that shows the amount of additional sales necessary to pay for selected levels of annual costs at certain profit margins. Duplications of this chart containing identical numbers appear in many places, as does the following example. The source of the data in Table 6 is a Federal Motor Carrier Safety Administration (FMCSA) publication, "Accident Cost Table." It was chosen because it is accessible online.

Table 6 presents a selection of the yearly accident cost levels shown in the FMCSA publication, but it still presents the idea adequately. To determine the additional sales necessary, accident costs are divided by the profit margin percentage selected, then converted into a decimal. In the FMCSA publication, accident costs are a combination of the indirect and direct costs.

Yearly accident costs	Profit margin				
	1%	2%	3%	4%	5%
$5,000	500,000	250,000	167,000	125,000	100,000
10,000	1,000,000	500,000	333,000	250,000	200,000
50,000	5,000,000	2,500,000	1,667,000	1,250,000	1,000,000
150,000	15,000,000	7,500,000	5,000,000	3,750,000	3,000,000
200,000	20,000,000	10,000,000	6,666,000	5,000,000	4,000,000

Table 6: Accident Cost Table
This table shows the dollars of revenue required to pay for different amounts of costs of accidents. It is necessary for a motor carrier to generate an additional $1.25 million revenue to pay the cost of a $25,000 accident, assuming an average profit of 2%. The amount of revenue required to pay for losses will vary with the profit margin. Note. Data from "Accident Cost Table," by Federal Motor Carrier Safety Administration, Washington, DC: Author.

OSHA's $afety Pays program is another example of how Bird's concept has been applied. The program literature says the program will, among other things, allow the user to "generate a report of the costs and the sales needed to cover those costs." As computations required by the program are made, a profit percentage is to be entered to produce the amount of sales necessary to cover direct and indirect costs.

As noted, statistics comparable to those in Table 6 were discussed with an operations executive whose degree is in finance. This conversation was highly instructive, particularly with respect to unit pricing methods and break-even charts, and why the computations to determine the additional sales necessary to cover indirect and direct costs are implausible. The discussion, briefly summarized, follows.

- As a company budgets for its operations, both the indirect and direct costs of injuries (unidentified for indirect costs and sometimes identified for direct costs) are included in estimates of operating costs, and the unit prices for products or services are set to recover those costs and the profit margin expected.
  - Thus, from the first dollar of sales onward, a part of the indirect and direct costs are recovered.
  - When revenues equal total operating costs (Figure 2), all indirect and direct costs incurred up to that time are recovered. Also, as sales increase and operations become profitable, revenues obtained continue to encompass all additional indirect and direct costs.
  - No "additional sales" are needed to cover indirect and direct costs.

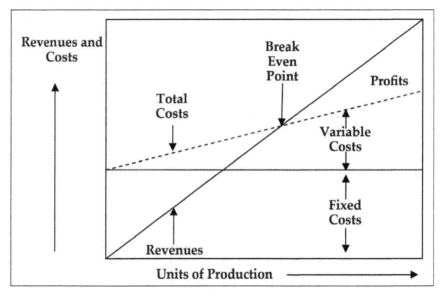

Figure 2: Break-Even Chart

Consider the following example:

1) Sales budgeted for a year are $10 million.

2) A 3% profit margin and profits of $300,000 are expected.

3) Operating costs are $9.7 million.

4) Total direct and indirect costs are $300,000.

5) To determine the additional sales necessary to cover total injury costs, the $300,000 would be divided by 0.03, the result being $10 million.

Such computations will not withstand a logic test, since all indirect and direct injury costs would be contained in the operating costs as the $10 million sales goal is attained. Thus, no additional sales, in addition to the organization's total income, are necessary to cover accident costs.

Conclusion

The literature on direct and indirect costs does not present a uniformly accepted computation method. Differences in the various systems are substantial. More importantly, no published ratios are currently valid because the increase in direct costs (indemnity and medical costs resulting from an injury or illness) has exceeded the increase in indirect costs substantially in the past 15 years.

Computations in this article updating the Stanford study indicate that the ratio of indirect to direct accident costs is currently about 0.8:1. That ratio is given as an approximation. Safety professionals who use a 1-to-1 ratio can be reasonably comfortable. This author recommends avoiding the use of ratios for which there are no supporting data (4:1, 6:1, 10:1 or higher).

Petersen (1989) also expressed concern over the use of an indirect to direct cost ratio for which supporting data are questionable:

> Although hidden costs are very real, they are very difficult to demonstrate. To say arbitrarily to management that they amount to four times the insurable costs is asking for trouble. If management asks for proof, you can only say, "Heinrich said so." Management wants facts, not fantasy. Without proof, hidden costs become fantasy.

SH&E professionals need proof to support the indirect to direct cost ratios they use. ASSE or its Foundation should consider funding research to develop valid and current data on indirect to direct costs ratio. For such a study to be successful, the methodology and scope should follow good research protocols, and a statistically based number of employers would need to be educated on and committed to the time and effort necessary. Such studies are not easily undertaken. That said, such a study will likely result in a determination that the indirect to direct cost ratio is significantly lower than the ratios that have appeared in safety literature. ∎

References

Bird, F.E. Jr. (1974). *Management guide to loss control*. Loganville, GA: Institute Press.

Bird, F.E. Jr. & Germain, G.L. (1985) *Practical loss control leadership* (Revised ed.). Loganville, GA: Det Norske Veritas.

Canadian Manufacturers and Exporters (Ontario Division) & Workplace Safety and Insurance Board. (2001). *The business results through health and safety guidebook*. Toronto, ON: Author. Retrieved Nov. 11, 2010, from www.wsib.on.ca/wsib/wsibsite.nsf/public/BusinessResults-HealthSafety.

Choi, S.D. (2006, Spring). A survey of the safety roles and costs of injuries in the roofing contracting industry. *Journal of SH&E Research, 3*(1), 1-20. Retrieved Nov. 11, 2010, from www.asse.org/academicsjournal/archive/vol3no1/06spring_choi5.pdf.

Federal Motor Carrier Safety Administration. Accident cost table. Washington, DC: Author. Retrieved Nov. 11, 2010, from www.fmcsa.dot.gov/facts-research/facts-figures/analysis-statistics/Revenue.htm.

Grimaldi, J.V. & Simonds, R.H. (1989). *Safety Management* (5th ed.). Homewood, IL: Irwin Press.

Heinrich, H.W. [1931; 1959 (4th ed.)]. *Industrial accident prevention: A scientific approach*. New York: McGraw-Hill Book Co.

International Labor Organization. Introduction to occupational safety and health (Training module). Geneva, Switzerland: Author.

International Safety Equipment Association (ISEA). (2002, March). Personal protective equipment: An investment in your workers' and company's future. Arlington, VA: Author.

Leigh, J.P., Markowitz, S.B., Fahs, M.C., et al. (1997). Costs of occupational injuries and illnesses in 1992. Washington, DC: U.S. Department of Health and Human Services, CDC, NIOSH.

Mealy, D.C. (2009). 2009 state of the line. Boca Raton, FL: National Council on Compensation Insurance. Retrieved Nov. 11, 2010, from www.ncci.com/Documents/AIS-09-SOL-Article.pdf.

OSHA. Do you know how much accidents are really costing your business? Washington, DC: U.S. Department of Labor (DOL), Author. Retrieved Nov. 11, 2010, from www.osha.gov/SLTC/etools/safetyhealth/images/safpay1.gif

OSHA. (2007). Safety and health management systems eTool: Costs of accidents. Washington, DC: U.S. DOL, Author. Retrieved Nov. 11, 2010, from www.osha.gov/SLTC/etools/safetyhealth/mod1_costs.html.

OSHA. (2009a). $afety Pays. Washington, DC: U.S. DOL, Author. Retrieved Nov. 11, 2010, from www.osha.gov/dcsp/smallbusiness/safetypays/index.html.

OSHA. (2009b). $afety Pays: Background of the cost estimates. Washington, DC: U.S. DOL, Author. Retrieved Nov. 11, 2010, from www.osha.gov/dcsp/smallbusiness/safetypays/background.html.

OSHA. (2009c). $afety Pays: Estimated costs of occupational injuries and illnesses and estimated impact on a company's profitability worksheet. Washington, DC: U.S. DOL, Author. Retrieved Nov. 11, 2010, from www.osha.gov/dcsp/smallbusiness/safetypays/estimator.html.

Petersen, D. (1989). *Techniques of safety management* (3rd ed.). Goshen, NY: Aloray Inc.

Stanford University. (1981). Improving construction safety performance: The user's role (Technical report No. 260.). Palo Alto, CA: Author, Department of Civil Engineering.

The Business Roundtable. (1982). Improving construction safety performance (Report A-3). New York: Author.

U.S. Fish & Wildlife Service. Definitions. Arlington, VA: Author, Division of Safety and Health. Retrieved Nov. 11, 2010, from www.fws.gov/safety/definitions.htm.

Western National Insurance. The total costs of accidents and how they affect your profits. Retrieved Nov. 11, 2010, from www.wnins.com/resources/commercial/Spotlight/Total%20cost%20of%20accidents.pdf.

10 Reviewing Heinrich
Dislodging Two Myths from the Practice of Safety

Originally published October 2011

In *The Standardization of Error*, Stefansson (1928) makes the case that people are willing to accept as fact what is written or spoken without adequate supporting evidence. When studies show that a supposed fact is not true, dislodging it is difficult because that belief as become deeply embedded in the minds of people and, thereby, standardized.

Stefansson pleads for a mind-set that accepts as knowledge only that which can be proven and which cannot be logically contradicted. He states that his theme applies to all fields of endeavor except for mathematics. Safety is a professional specialty in which myths have become standardized and deeply embedded. This article examines two myths that should be dislodged from the practice of safety:

1) Unsafe acts of workers are the principal causes of occupational accidents.
2) Reducing accident frequency will equivalently reduce severe injuries.

These myths arise from the work of H.W. Heinrich (1931; 1941; 1950; 1959). They can be found in the four editions of *Industrial Accident Prevention: A Scientific Approach*. Although some safety practitioners may not recognize Heinrich's name, his misleading premises are perpetuated as they are frequently cited in speeches and papers.

Analytical evidence indicates that these premises are not soundly based, supportable or valid, and, therefore, must be dislodged. Although this article questions the validity of the work of an author whose writings have been the foundation of safety-related teaching and practice for many decades, it is appropriate to recognize the positive effects of his work as well.

This article was written as a result of encouragement from several colleagues who encountered situations in which these premises were cited as fact, with the resulting recommended preventive actions being inappropriate and ineffective. Safety professionals must do more to inform about and refute these myths so that they may be dislodged.

Recognition: Heinrich's Achievements

Heinrich was a pioneer in the field of accident prevention and must be given his due. Publication of his book's four editions spanned nearly 30 years. From the 1930s to today, Heinrich likely has had more influence than any other individual on the work of occupational safety practitioners. In retrospect, knowing the good done by him in promoting greater attention to occupational safety and health should be balanced with an awareness of the misdirection that has resulted from applying some of his premises.

Citation	1st 1931	2nd 1941	3rd 1950	4th 1959
H-1	128	269	32	17
H-2	127	268	325	173
H-3				171
H-4	127	268	325	173
H-5	45	19	18	21
H-6	128	269	326	174
H-7		22	2	4
H-8		20	18	21
H-9	44	20	18	20
H-10	95	34	31	33
H-11	31			
H-12		26		
H-13	91			
H-14		27	24	27
H-15		27	24	27
H-16				31
H-17		26		
H-18			24	26
H-19			24	26
H-20	101			
H-21			188	
H-22	22			28

Table 1: Pages Cited by Edition

Heinrich's Sources Unavailable

Attempts were made to locate Heinrich's research, without success. Dan Petersen, who with Nestor Roos, authored a fifth edition of *Industrial Accident Prevention*, was asked whether they had located Heinrich's research. Petersen said that they had to rely entirely on the previous editions of Heinrich's books as resources. Thus, the only data that can be reviewed are contained in Heinrich's books. His information-gathering methods, survey documents that may have been used, the quality of the information gathered and the analytical systems used cannot be examined.

Two items of note for this article: Citations from Heinrich's texts are numbered H-1, H-2, etc., and correspond to the chart in Table 1, which indicates the page numbers and editions in which each citation appears. All other citations appear as in-text references in the journal's standard style.

Furthermore, in today's social climate, some of Heinrich's terminology would be considered sexist. He uses phrases such as *man failure*, *the foreman* and *he is responsible*. Consider the time in which he wrote. The fourth edition was published in 1959.

Psychology and Safety

Applied psychology dominates Heinrich's work with respect to selecting causal factors and is given great importance in safety-related problem resolution. Consider the following:

1) Heinrich expresses the belief that "psychology in accident prevention is a fundamental of great importance" (H-1).

2) His premise is that "psychology lies at the root of sequence of accident causes" (H-2).

3) In the fourth edition, Heinrich states that he envisions "the more general acceptance by management of the idea that an industrial psychologist be included as a member of the plant staff as a physician is already so included" (H-3).

4) The focus of applied psychology on the employee, as in the following quotation:

> Indeed, safety psychology is as fairly applicable to the employer as to the employee. The initiative and the chief burden of activity in accident prevention rest upon the employer; however the practical field of effort for prevention through psychology is confined to the employee, but through management and supervision. (H-4)

Note that the focus of applied psychology is on the worker as are other Heinrichean premises. Since application of practical psychology is confined to the worker, who reports to a supervisor, the psychology applier is the supervisor. With due respect to managers, supervisors and safety practitioners, it is doubtful that many could knowledgeably apply psychology "as a fundamental of great importance" in their accident prevention efforts.

Heinrich's Causation Theory: The 88-10-2 Ratio

Heinrich professes that among the direct and proximate causes of industrial accidents:
- 88% are unsafe acts of persons;
- 10% are unsafe mechanical or physical conditions;
- 2% are unpreventable (H-5).

According to Heinrich, man failure is the problem and psychology is an important element in correcting it. In his discussion of the relation of psychology to accident prevention, Heinrich advocates identifying the first proximate and most easily prevented cause in the selection of remedies. He says:

> Selection of remedies is based on practical cause-analysis that stops at the selection of the first proximate and most easily prevented cause (such procedure is advocated in this book) and considers psychology when results are not produced by simpler analysis. (H-6)

Note that the first proximate and most easily prevented cause is to be selected (88% of the time a human error). That concept permeates Heinrich's work. It does not encompass what has been learned subsequently about the complexity of accident causation or that other causal factors may be more significant than the first proximate cause.

For example, the *Columbia* Accident Investigation Board (NASA, 2003) notes the need to consider the complexity of incident causation:

> Many accident investigations do not go far enough. They identify the technical cause of the accident, and then connect it to a variant of "operator error." But this is seldom the entire issue. When the determinations of the causal chain are limited to the technical flaw and individual failure, typically the actions taken to prevent a similar event in the future are also limited: fix the technical problem and replace or retrain the individual responsible.
>
> Putting these corrections in place leads to another mistake: The belief that the problem is solved. Too often, accident investigations blame a failure only on the last step in a complex process, when a more comprehensive understanding of that process could reveal that earlier steps might be equally or even more culpable.

A recent example of the complexity of accident causation appears in this excerpt from the report prepared by BP personnel following the April 20, 2010, *Deepwater Horizon* explosion in the Gulf of Mexico (BP, 2010):

> The team did not identify any single action or inaction that caused this incident. Rather, a complex and interlinked series of mechanical failures, human judgments, engineering

design, operational implementation and team interfaces came together to allow the initiation and escalation of the accident.

Consider another real-world situation in which a fatality resulted from multiple causal factors:

> An operation produces an odorless, colorless highly toxic gas in an enclosed area. The two-level gas detection and alarm system has deteriorated over many years of use, and the system often leaks gas. An internal auditor recommends it be replaced with a three-level system, the accepted practice in the industry for that type of gas. The auditor also recommends that maintenance give the existing system high priority.
>
> Management puts high profits above safety and tolerates excessive risk taking. That defines culture problems. Management decides not to replace the system, and furthermore begins a cost-cutting initiative that reduces maintenance staff by one-third. The gas detection and alarm system continue to deteriorate, and maintenance staff cannot keep up with the frequent calls for repair and adjustment.
>
> A procedure is installed that requires employees to test for gas before entering the enclosed area. But, supervisors condone employees entering the area without making the required test. Both detection and alarm systems fail. Gas accumulates. An employee enters the area without testing for gas. The result is a toxic gas fatality.
>
> Causal factor determination would commence with the deficiencies in the organization's culture whereby: resources were not provided to replace a defective detection and alarm system in a critical area; staffing decisions resulted in inadequate maintenance; and excessive risk taking was condoned. The employee's violation of the established procedure was a contributing factor, but not principal among several factors.

Heinrich's theory that an unsafe act is the sole cause of an accident is not supported in the cited examples. Also, note that Heinrich's focus on man failure is singular in the following citation: "In the occurrence of accidental injury, it is apparent that man failure is the heart of the problem; equally apparent is the conclusion that methods of control must be directed toward man failure" (H-7). [Note: Heinrich does not define man failure. In making the case to support directing efforts toward controlling man failure, he cites personal factors such as unsafe acts, using unsafe tools and willful disregard of instruction.]

A directly opposite view is expressed by Deming (1986). Deming is known for his work in quality principles, which this author finds comparable to the principles required to achieve superior results in safety.

> The supposition is prevalent throughout the world that there would be no problems in production or service if only our production workers would do their jobs in the way that we taught. Pleasant dreams. The workers are handicapped by the system, and the system belongs to the management. (p. 134)

Of all Heinrich's concepts, his thoughts on accident causation, expressed as the 88-10-2 ratios, have had a significant effect on the practice of safety, and have resulted in the most misdirection. Why is this so? Because when based on the premise that man failure causes the most accidents, preventive efforts are directed at the worker rather than toward the operating system in which the work is performed.

Many safety practitioners operate on the belief that the 88-10-2 ratios are soundly based and, as a result, focus their efforts on reducing so-called man failure rather than attempting to improve the system. This belief also perpetuates because it is the path of least resistance for an organization. It is easier for supervisors and managers to be satisfied with taking superficial preventive action, such as retraining a worker, reinstructing the work group or reposting the standard operating procedure, than it is to try to correct system problems.

Certainly, operator errors may be causal factors for accidents. However, consider Ferry's (1981) comments on this subject:

> We cannot argue with the thought that when an operator commits an unsafe act, leading to a mishap, there is an element of human or operator error. We are, however, decades past the place where we once stopped in our search for causes.
>
> Whenever an act is considered unsafe we must ask why. Why was the unsafe act committed? When this question is answered in depth it will lead us on a trail seldom of the operator's own conscious choosing. (p. 56)

If, during an accident investigation, a professional search is made for causal factors beyond an unsafe act, such as through the five-why method, one will likely find that the causal factors built into work systems may be of greater importance than an employee's unsafe act. Fortunately, a body of literature has emerged that recognizes the significance of causal factors which originate from decisions made above the worker level. Several are cited here.

Human Errors Above the Worker Level

Much as been written about human error. Particular attention is given to the *Guidelines for Preventing Human Error in Process Safety* (CCPS, 1994). Although process safety appears in the title, the first two chapters provide an easily read primer on human error reduction. The content of those chapters was largely influenced by personnel with plant- or corporate-level safety management experience.

Safety practitioners should view the following highlights as generic and broadly applicable. They advise on where human errors occur, who commits them and at what level, the effect of organizational culture and where attention is needed to reduce the occurrence of human errors. These highlights apply to organizations of all types and sizes.

- It is readily acknowledged that human errors at the operational level are a primary contributor to the failure of systems. It is often not recognized, however, that these errors frequently arise from failures at the management, design or technical expert levels of the company (p. xiii).
- A systems perspective is taken that views error as a natural consequence of a mismatch between human capabilities and demands, and an inappropriate organizational culture. From this perspective, the factors that directly influence error are ultimately controllable by management (p. 3).
- Almost all major accident investigations in recent years have shown that human error was a significant causal factor at the level of design, operations, maintenance or the management process (p. 5).
- One central principle presented in this book is the need to consider the organizational factors that create the preconditions for errors, as well as the immediate causes (p. 5).
- Attitudes toward blame will determine whether an organization develops a blame culture, which attributes error to causes such as lack of motivation or deliberate unsafe behavior (p. 5).
- Factors such as the degree of participation that is encouraged in an organization, and the quality of the communication between different levels of management and the workforce, will have a major effect on the safety culture (p. 5).

Since "failures at the management, design or technical expert levels of the company" affect the design of the workplace and the work methods—that is, the operating system—it is logical to suggest that safety professionals should focus on system improvement to attain acceptable risk levels rather than principally on affecting worker behavior.

Reason's (1997) book, *Managing the Risks of Organizational Accidents*, is a must-read for safety professionals who want an education in human error reduction. It has had five additional printings since 1997. Reason writes about how the effects of decisions accumulate over time and become the causal factors for incidents resulting in serious injuries or major damage when all the circumstances necessary for the occurrence of a major event fit together. This book stresses the need to focus on decision making above the worker level to prevent major accidents. Reason states:

> Latent conditions, such as poor design, gaps in supervision, undetected manufacturing defects or maintenance failures, unworkable procedures, clumsy automation, shortfalls in training, less than adequate tools and equipment, may be present for many years before they combine with local circumstances and active failures to penetrate the system's layers of defenses.
>
> They arise from strategic and other top-level decisions made by governments, regulators, manufacturers, designers and organizational managers. The impact of these deci-

sions spreads throughout the organization, shaping a distinctive corporate culture and creating error-producing factors within the individual workplaces. (p. 10)

The traditional occupational safety approach alone, directed largely at the unsafe acts of persons, has limited value with respect to the "insidious accumulation of latent conditions [that he notes are] typically present when organizational accidents occur. (pp. 224, 239)

If the decisions made by management and others have a negative effect on an organization's culture and create error-producing factors in the workplace, focusing on reducing human errors at the worker level—the unsafe acts—will not address the problems.

Deming achieved world renown in quality assurance. The principle embodied in what is referred to as Deming's 85-15 rule also applies to safety. The rule supports the premise that prevention efforts should be focused on the system rather than on the worker. This author draws a comparable conclusion as a result of reviewing more than 1,700 incident investigation reports. This is the rule, as cited by Walton (1986): "The rule holds that 85% of the problems in any operation are within the system and are the responsibly of management, while only 15% lie with the worker" (p. 242).

In 2010, ASSE sponsored the symposium, Rethink Safety: A New View of Human Error and Workplace Safety. Several speakers proposed that the first course of action to prevent human errors is to examine the design of the work system and work methods. Those statements support Deming's 85-15 rule. Consider this statement by a human error specialist [from this author's notes]:

When errors occur, they expose weaknesses in the defenses designed into systems, processes, procedures and the culture. It is management's responsibility to anticipate errors and to have systems and work methods designed so as to reduce error potential and to minimize severity of injury potential when errors occur.

Since most problems in an operation are systemic, safety efforts should be directed toward improving the system. Unfortunately, the use of the terms *unsafe acts* and *unsafe conditions* focuses attention on a worker or a condition, and diverts attention from the root-causal factors built into an operation.

Allied to Deming's view is the work of Chapanis, who was prominent in the field of ergonomics and human factors engineering. Representative of Chapanis's writings is "The Error-Provocative Situation," a chapter in *The Measurement of Safety Performance* (Tarrants, 1980). Chapanis's message is that if the design of the work is error-provocative, one can be certain that errors will occur in the form of accident causal factors. It is illogical to conclude in an incident investigation that the principal causal factor is the worker's unsafe act if the design of the workplace or the work methods is error-inviting. In such cases, the error-producing aspects of the work (e.g., design, layout, equipment, operations, the system) should be considered primary.

U.S. Department of Energy (1994) describes the management oversight and risk tree (MORT) as a "comprehensive analytical procedure that provides a disciplined method for determining the systemic causes and contributing factors of accidents." The following reference to "performance errors" is of particular interest.

> It should be pointed out that the kinds of questions raised by MORT are directed at systemic and procedural problems. The experience, to date, shows there are a few "unsafe acts" in the sense of blameful work level employee failures. Assignment of "unsafe act" responsibility to a work-level employee should not be made unless or until the preventive steps of 1) hazard analysis; 2) management or supervisory direction; and 3) procedures safety review have been shown to be adequate. (p. 19)

Each of these more recent publications refutes the premise that unsafe acts are the primary causes of occupational accidents.

Heinrich's Data Gathering and Analytical Method

Heinrich recognized that other studies on accident causation identified both unsafe acts and unsafe conditions as causal factors with almost equal frequency. Those studies produced results different from the 88-10-2 ratios. For example, the *Accident Prevention Manual for Industrial Operations: Administration and Programs*, 8th edition (NSC, 1980), contains these statements about studies of accident causation:

> Two historical studies are usually cited to pinpoint the contributing factor(s) to an accident. Both emphasize that most accidents have multiple causes.
> •A study of 91,773 cases reported in Pennsylvania in 1953 showed 92% of all nonfatal injuries and 94% of all fatal injuries were due to hazardous mechanical or physical conditions. In turn, unsafe acts reported in work injury accidents accounted for 93% of the nonfatal injuries and 97% of the fatalities.
> •In almost 80,000 work injuries reported in that same state in 1960, unsafe condition(s) was identified as a contributing factor in 98.4% of the nonfatal manufacturing cases, and unsafe act(s) was identified as a contributing factor in 98.2% of the nonfatal cases. (p. 241)

Although aware that others studying accident causation had recognized the multifactorial nature of causes, Heinrich continued to justify selecting a single causal factor in his analytical process. Heinrich's data-gathering methods force the accident cause determination into a singular and narrow categorization. The following paragraph is found in the second through fourth editions. It follows an explanation of the study resulting in the formulation of the 88-10-2 ratios. "In this research, major responsibility for each accident was assigned either to the unsafe act of a person or to an unsafe mechanical condition, but in no case were both personal and mechanical causes charged" (H-8).

Heinrich's study resulting in the 88-10-2 ratios was made in the late 1920s. Both the relationship of a study made then to the work world as it now exists and the methods used in producing it are questionable and unknown. As to the study methods, consider the following paragraph, which appears in the first edition; minor revisions were made in later editions.

> Twelve thousand cases were taken at random from closed-claim-file insurance records. They covered a wide spread of territory and a great variety of industrial classifications. Sixty-three thousand other cases were taken from the records of plant owners. (H-9)

The source of the data was insurance claims files and records of plant owners, which cannot provide reliable accident causal data because they rarely include causal factors. Nor are accident investigation reports completed by supervisors adequate resources for causal data. When this author provided counsel to clients in the early stages of developing computer-based incident analysis systems, insurance claims reports and supervisors' investigation reports were examined as possible sources for causal data. It was rare for insurance claims reports to include provisions to enter causal data.

This author has examined more than 1,700 incident investigation reports completed by supervisors and investigation teams. In approximately 80% of those reports, causal factor information was inadequate. These reports are not a sound base from which to analyze and conclude with respect to the reality of causal factors.

Summation on the 88-10-2 Ratios

Heinrich's data collection and analytical methods in developing the 88-10-2 ratios are unsupportable. Heinrich's premise, that unsafe acts are the primary causes of occupational accidents, cannot be sustained. The myth represented by those ratios must be dislodged and actively refuted by safety professionals.

An interesting message of support with respect to avoiding use of the 88-10-2 ratios comes from Krause (2005), a major player in worker-focused behavior-based safety:

> Many in the safety community believe a high percentage of incidents, perhaps 80% to 90%, result from behavioral causes, while the remainder relate to equipment and facilities. We made this statement in our first book in 1990. However, we now recognize that this dichotomy of causes, while ingrained in our culture generally and in large parts of the safety community, is not useful, and in fact can be harmful. (p. 10)

The Foundation of a Major Injury: The 300-29-1 Ratios

Heinrich's conclusion with respect to the ratios of incidents that result in no injuries, minor injuries and a major lost-time case was the base on which educators taught and many safety practitioners came to believe that reducing accident frequency will achieve equivalent reduction in injury severity. The following statement appears in all four editions of his text: "The natural conclusion follows, moreover, that in the largest injury group—the minor injuries—lies the most valuable clues to accident causes" (H-10).

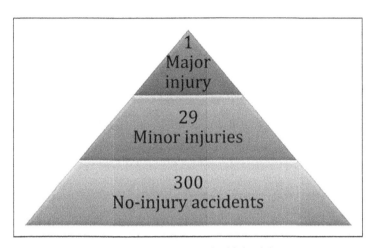

Figure 1: Foundation of a Major Injury
Note. Adapted from *Industrial Accident Prevention: A Scientific Approach* (1st ed.) (p. 91), (2nd ed.) (p. 27), (3rd ed.) (p. 24), (4th ed.) (p. 27), by H.W. Heinrich, 1931, 1941, 1950, 1959, New York: McGraw-Hill.

The following discussion and statistics establish that the ratios upon which the foregoing citation is based are questionable and that reducing incident frequency does not necessarily achieve an equivalent reduction in injury severity. Heinrich's 300-29-1 ratios have been depicted as a triangle or a pyramid (Figure 1). In his first edition, Heinrich writes:

> Analysis proves that for every mishap resulting in an injury there are many other accidents in industry which cause no injuries whatever. From data now available concerning the frequency of potential-injury accidents, it is estimated that in a unit group of 330 accidents, 300 result in no injuries, 29 in minor injuries, and 1 in a major or lost-time case. (H-11)

In the second edition, "similar" was added to the citation: "Analysis proves that for every mishap, there are many other similar accidents in industry . . ." (H-12).

Within a chart displaying the 300-29-1 ratios in the first edition, Heinrich writes, "The total of 330 accidents all have the same cause." Note that cause is singular (H-13). This statement, that all 330 incidents have the same cause, challenges credulity. Also, note that the sentence quoted in this paragraph appears only in the first edition. It does not appear in later editions (H-14).

For background data, Heinrich says in the first, second and third editions:

> The determination of this no-injury accident frequency followed a *most interesting and absorbing study* [italics added]. The difficulties can be readily imagined. There were few existing data on minor injuries—to say nothing of no-injury accidents. (H-15)

In the fourth edition, published 28 years after the first edition, the source of the data is more specifically stated:

> The determination of this no-injury accident frequency followed a *study of over 5,000 cases* [italics added]. The difficulties can be readily imagined. There were few existing data on minor injuries—to say nothing of no-injury accidents. (H-16)

The credibility of such a revision after 28 years must be questioned. In Heinrich's second and third editions, major changes were made in his presentation on the ratios, without explanation.

1) The statement in the first edition that the 330 accidents all have the same cause was eliminated.

2) In the second edition, changes were made indicating that the unit group of 330 accidents are "similar" and "of the same kind" (H-17).

3) In the third edition, another significant addition is made. The 330 accidents now are "of the same kind and involving the same person" (H-18).

The following appears in the third and fourth editions, encompassing the changes noted.

> Analysis proves that for every mishap resulting in an injury there are many other similar accidents that cause no injuries whatever. From data now available concerning the frequency of potential-injury accidents, it is estimated that in a unit group of 330 accidents *of the same kind and involving the same person* [italics added], 300 result in no injuries, 29 in minor injuries and 1 in a major or lost-time injury. (H-19)

These changes are not explained. If the original data were valid, how does one explain the substantial revisions in the conclusions eventually drawn from an analysis of it? In the second, third and fourth editions, Heinrich gives no indication of other data collection activities or of other analyses. How does one support using the ratios without having explanations of the differing interpretations Heinrich gives in each edition?

The changes made in the 300-29-1 ratios in the second and third editions, and carried over into the fourth edition, present other serious conceptual problems. To which types of accidents does "in a unit group of 330 accidents of the same kind and occurring to the same person" apply? Certainly, it does not apply to some commonly cited incident types, such as falling to a lower level or struck by objects.

For example, a construction worker rides the hoist to the 10th floor and within minutes backs into an unguarded floor opening, falling to his death. Heinrich's ratios would give this person favorable odds of 300 to 330 (10 out of 11) of suffering no injury at all. That is not credible.

Consider the feasibility of finding data in the 5,000-plus cases studied to support the ratios, keeping in mind that incidents are to be of the same type and occurring to the same person.

- If the number of major or lost-time cases is 1, the number of minor injury case files would be 29 and the number of no-injury case files would be 300.
- If the number of major or lost-time cases is 5, the number of minor injury case files would be 145 and the number of no-injury case files would be 1,500.
- If the number of major or lost-time cases is 10, the number of minor injury case files would be 290 and the number of no-injury case files would be 3,000.

Because of the limitations Heinrich himself imposes, that all incidents are to be of the same type and occurring to the same person, it is implausible that his database could contain the information necessary for analysis and the conclusions he drew on his ratios. Particularly disconcerting is the need for the database to contain information on more than 4,500 no-injury cases (300 ÷ 330 × 5,000). Unless a special study was initiated, creating files on no-injury incidents would be a rarity.

Given this, one must ask, did a database exist upon which Heinrich established his ratios, then stated the premises that the most valuable clues for accident causes are found in the minor injury category? This author thinks not.

Statistical Indicators: Serious Injury Trending

Data on the trending of serious injuries and workers' compensation claims contradict the premise that focusing on incident frequency reduction will equivalently achieve severity reduction. The following data have been extracted from publications of the National Council on Compensation Insurance (NCCI, 2005; 2006; 2009).

- In 2006, NCCI produced a 12-minute video, *The Remarkable Story of Declining Frequency—Down 30% in the Past Decade*. It shows that workers' compensation claim frequency was down considerably in the decade cited. The video tells a remarkable but not well-known story.
- A July 2009 NCCI bulletin, "Workers' Compensation Claim Frequency Continues Its Decline in 2008." The reduction was 4.0%. A May 2010 NCCI report says that the cumulative reduction in claims frequency from 1991 through 2008 is 54.7%.

• A 2005 NCCI paper, "Workers' Compensation Claim Frequency Down Again," states, "There has been a larger decline in the frequency of smaller lost-time claims than in the frequency of larger lost-time claims." Also, consider that NCCI (2005) reports reductions in selected categories of claim values for the years 1999 and 2003, expressed in 2003 hard dollars (Table 2).

Value of claim	Declines in frequency
Less than $2,000	34%
$2,000 to $10,000	21%
$10,000 to $50,000	11%
More than $50,000	7%

Table 2: Injury Reduction Categories
Note. Data from "State of the Line," by National Council on Compensation Insurance, 2005, Boca Raton, FL: Author.

While the frequency of workers' compensation cases is down, the greatest reductions are for less serious injuries. The reduction in cases valued from $10,000 to $50,000 is about one-third of that for cases valued at less than $2,000. For cases valued above $50,000, the reduction is about one-fifth of that for the less costly and less serious injuries. The data clearly show that a comparable reduction in injury severity does not follow a reduction in injury frequency.

A DNV (2004) bulletin is another resource of particular note. It states that managing operations to reduce frequency will not equivalently reduce severity.

What about the pyramid?
Much has been said over the years about the classical loss control pyramid, which indicates the ratio between no loss incidents, minor incidents and major incidents, and it has often been argued that if you look after the small potential incidents, the major loss incidents will improve also.

The major accident reality however is somewhat different. What we find is that if you manage the small incidents effectively, the small incident rate improves, but the major accident rate stays the same, or even slightly increases.

Contradictions: Unsafe Acts and Injuries

Heinrich's texts contain contradictions about when a major injury would occur and the relationship between unsafe acts and a major injury. In all editions, reference is made to 330 careless acts or several hundred unsafe acts occurring before a major injury occurs, as in the following examples from the first and third editions.

- "Keep in mind that a careless act occurs approximately 300 times *before* [italics added] a serious injury results and that there is, therefore, an excellent opportunity to detect and correct unsafe practices before injury occurs" (H-20).

- "Keep in mind that an unsafe act occurs several hundred times *before* [italics added] a serious injury results" (H-21).

Before is a key word here. While an unsafe act may be performed several times before a particular accident occurs, that is not the case in a large majority of incidents which result in serious injury or fatality. In his fourth edition, Heinrich gave this view of the relationship of unsafe acts or exposures to mechanical hazards.

> If it were practicable to carry on appropriate research, still another base therefore could be established showing that from 500 to 1,000 or more unsafe acts or exposures to mechanical hazards existed in the average case before even one of the 300 narrow escapes from injury (events-accidents) occurred. (H-22)

There is a real problem here. All of those unsafe acts or exposures to mechanical hazards take place before even one accident occurs. That is illogical.

Summation on the 300-29-1 Ratios

Use of the 300-29-1 ratios is troubling. Since the ratios are not soundly based, one must ask whether the ratios have any substance. Does their use as a base for a safety management system result in a concentration of resources on the frequent and lesser significant while ignoring opportunities to reduce the more serious injuries?

One of Heinrich's premises is that "the predominant causes of no-injury accidents are, in average cases, identical with the predominant causes of major injuries, and incidentally of minor injuries as well." This is wrong. It is a myth that must be dislodged from the practice of safety.

Applying this premise leads to misdirection in resource application and ineffectiveness, particularly with respect to preventing serious injuries. In this author's experience, many incidents resulting in serious injury are singular and unique events, with multifaceted and complex causal factors, and descriptions of similar incidents are rare in the historical body of incident data. Furthermore, all hazards do not have equal potential for harm. Some risks are more significant than others. That requires priority setting.

Misinterpretation of Terms

Not only have many safety practitioners used the 300-29-1 ratios in statistical presentations, but many also have misconstrued what Heinrich intended with the terms *major injury, minor injury* and *no-injury accidents*. Some practitioners who cite these ratios in their presentations assume that a "major injury" is a serious injury or a fatality. In each edition, Heinrich gave nearly identical definitions of the accident categories to which the 300-29-1 ratios apply. This is how the definition reads in the fourth edition.

> In the accident group (330 cases), a major injury is any case that is reported to insurance carriers or to the state compensation commissioner. A minor injury is a scratch, bruise or laceration such as is commonly termed a first-aid case. A no-injury accident is an unplanned event involving the movement of a person or an object, ray or substance (e.g., slip, fall, flying object, inhalation) having the probability of causing personal injury or property damage. The great majority of reported or major injuries are not fatalities or fractures or dismemberments; they are not all lost-time cases, and even those that are do not necessarily involve payment of compensation. (H- 20)

These definitions compel the conclusion that any injury requiring more than first-aid treatment is a major injury. When these definitions were developed in the late 1920s, few companies were self-insured for workers' compensation. On-site medical facilities were rare. Insurance companies typically paid for medical-only claims and for minor and major injuries. According to Heinrich's definitions, almost all such claims would be considered major injuries. Then, is it not so that every OSHA recordable injury is a major injury in this context?

Heinrich's 300-29-1 ratios have been misused and misrepresented many times as well. For example, a safety director recently said that in the previous year his company sustained one fatality and 30 OSHA days-away-from-work incidents, and, therefore, Heinrich's progression was validated. Not so. All of the injuries and the fatality would be in the major or lost-time injury category.

In another instance, a speaker referred to Heinrich's 300-29-1 ratios and said that the 300 were unsafe acts, the 29 were serious injuries and the 1 was a fatality. These are but two examples of the many misuses of these ratios.

Heinrich's Premises Versus Current Safety Knowledge

Heinrich emphasized improving an individual worker's performance, rather than improving the work system established by management. That is not compatible with current knowledge. Unfortunately, some safety practitioners continue to base their counsel on Heinrich's premises, which narrows the scope of their activities as they attempt principally to improve worker performance. In doing so, they ignore the knowledge that has evolved in the professional practice of safety. A few examples follow:

- Hazards are the generic base of, and the justification for the existence of, the practice of safety.
- Risk is an estimate of the probability of a hazard-related incident or exposure occurring and the severity of harm or damage that could result.
- The entirety of purpose of those responsible for safety, regardless of their titles, is to manage their endeavors with respect to hazards so that the risks deriving from those hazards are acceptable.
- All risks to which the practice of safety applies derive from hazards. There are no exceptions.
- Hazards and risks are most effectively and economically avoided, eliminated or controlled in the design and redesign processes.
- The professional practice of safety requires consideration of the two distinct aspects of risk:

1) avoiding, eliminating or reducing the probability of a hazard-related incident or exposure occurring;

2) reducing the severity of harm or damage if an incident or exposure occurs.

- Management creates the safety culture, whether positive or negative.
- An organization's culture, translated into a system of expected behavior, determines management's commitment or lack of commitment to safety and the level of safety achieved.
- Principal evidence of an organization's culture with respect to occupational risk management is demonstrated through the design decisions that determine the facilities, hardware, equipment, tooling, materials, processes, configuration and layout, work environment and work methods.
- Major improvements in safety will be achieved only if a culture change takes place, only if major changes occur in the system of expected behavior.
- While human errors may occur at the worker level, preconditions for the commission of such errors may derive from decisions made with respect to the workplace and work methods at the management, design, engineering or technical expert levels of an organization.
- Greater progress can be obtained with respect to safety by focusing on system improvement to achieve acceptable risk levels, rather than through modifying worker behavior.
- A large proportion of problems in an operation are systemic, deriving from the workplace and work methods created by management, and can be resolved only by management. Responsibility for only a relatively small remainder lies with the worker.

- While employees should be trained and empowered up to their capabilities and encouraged to make contributions with respect to hazard identification and analysis, and risk elimination or control, they should not be expected to do what they cannot do.
- Accidents usually result from multiple and interacting causal factors that may have organizational, cultural, technical or operational systems origins.
- If accident investigations do not relate to actual causal factors, corrective actions taken will be misdirected and ineffective.
- Causal factors for low-probability/high-consequence events are rarely represented in the analytical data on incidents that occur frequently, and the uniqueness of serious injury potential must be adequately addressed. However, accidents that occur frequently may be predictors of severity potential if a high energy source was present (e.g., operation of powered mobile equipment, electrical contacts).

As this list demonstrates, Heinrich's premises are not compatible with current knowledge.

Recommendations

Safety professionals should ensure that the Heinrich misconceptions discussed in this article are discarded by the profession. To achieve this, each safety professional should:

- Stop using or promoting the premises that unsafe acts are the primary causes of accidents and that focusing on reducing accident frequency will equivalently reduce injury severity.
- Actively dispel these premises in presentations, writings and discussions.
- Politely but firmly refute allegations by others who continue to promote the validity of these premises.
- Apply current methods that look beyond Heinrich's myths to determine true causal factors of accidents.

Conclusion

As knowledge has evolved about how accidents occur and their causal factors, the emphasis is now properly placed on improving the work system, rather than on worker behavior. As one colleague who is disturbed by safety professionals who reference Heinrich premises as fact, says, "It is borderline unethical on their part."

This article has reviewed the origin of certain premises that have been accepted as truisms by many educators and safety practitioners, and how they evolved and changed over time; it also examined their validity. The two premises discussed here are wrongly based and cannot be sustained by safety practitioners. The premises themselves and the methods used to establish them cannot withstand a logic test. They are myths that have become deeply embedded in the practice of safety and safety professionals must take action to dislodge them. ■

References
BP. (2010, Sept. 8). Deepwater Horizon accident investigation report. Houston, TX: Author. Retrieved Aug. 30, 2011, from www.bp.com/liveassets/bp_internet/globalbp/global bp_uk_english/incident_response/STAGING/local_assets/downloads_pdfs/Deep water_Horizon_Accident_Investigation_Report.pdf.
Center for Chemical Process Safety (CCPS). (1994). Guidelines for preventing human error in process safety. New York: Author.
Columbia Accident Investigation Board. (2003). Columbia accident investigation report, Vol. 1. Washington, DC: NASA. Retrieved Aug. 30, 2011, from www.nasa.gov/colum bia/home/CAIB_Vol1.html.
Deming, W.E. (1986). Out of the crisis. Cambridge, MA: Center for Advanced Engineering Study, Massachusetts Institute of Technology.
Det Norske Veritas (DNV) Consulting. (2004). Leading indicators for major accident hazards: An invitation to industry partners. Houston, TX: Author.
Ferry, T.S. (1981). Modern accident investigation and analysis: An executive guide. New York: John Wiley & Sons.
Heinrich, H.W. (1931, 1941, 1950, 1959). Industrial accident prevention: A scientific approach. New York: McGraw-Hill. (See Table 1 for specific references.)
Heinrich, H.W., Petersen, D. & Roos, N. Industrial accident prevention (5th ed.). New York: McGraw-Hill.
Krause, T.R. (2005). Leading with safety. Hoboken, NJ: John Wiley & Sons.
Manuele, F.A. (2002). Heinrich revisited: Truisms or myths. Itasca, IL: National Safety Council.
Manuele, F.A. (2003). On the practice of safety (3rd ed.). New York: John Wiley & Sons.
Manuele, F.A. (2008, Dec.). Serious injuries and fatalities: A call for a new focus on their prevention. Professional Safety, 53(12), 32-39.
National Council on Compensation Insurance (NCCI). (2005, May). State of the line. Boca Raton, FL: Author. Retrieved Aug. 30, 2011, from www.ncci.com/media/pdf/ SOL_2005.pdf.
NCCI. (2006, June). Workers' compensation claim frequency down again in 2005 [Research brief]. Boca Raton, FL: Author. Retrieved Aug. 30, 2011, from www.ncci.com/docu ments/research-brief-august06.pdf.
NCCI. (2006, Nov.). The remarkable story of declining frequency—down 30% in the past decade [Video]. Boca Raton, FL: Author. Retrieved Aug. 30, 2011, from www.ncci. com/nccimain/IndustryInformation/NCCIVideos/ArchivedArticles/Pages/video_de clining_frequency_11-06.aspx.

NCCI. (2009, July). Workers' compensation claim frequency continues its decline in 2008 [Research brief]. Boca Raton, FL: Author. Retrieved Aug. 30, 2011, from www.ncci.com/Documents/WorkersCompensationClaimFrequency2008.pdf.

NCCI. (2010, May). State of the line. Boca Raton, FL: Author. Retrieved Aug. 30, 2011, from www.ncci.com/Documents/AIS-2010-SOL-Presentation.pdf.

National Safety Council (NSC). (1980). Accident prevention manual for industrial operations: Administration and programs (8th ed.). Itasca, IL: Author.

Reason, J. (1997). Managing the risks of organizational accidents. London: Ashgate Publishing Co.

Stefansson, V. (1928). The standardization of error. London: K. Paul, Trench, Trubner & Co. Ltd.

Tarrants, W.E. (1980). The measurement of safety performance. New York: Garland Publishing Co.

U.S. Department of Energy. (1994). Guide to use of the management oversight and risk tree (SSDC-103). Washington DC: Author.

Walton, M. (1986). The Deming management method. New York: The Putnam Publishing Group.

Acknowledgments

Parts of this article are updated material from three of the author's works: *Heinrich Revisited: Truisms or Myths*; chapter seven in *On the Practice of Safety* (3rd ed.); and the article, "Serious Injuries and Fatalities: A Call for a New Focus on Their Prevention," from the December 2008 issue of *Professional Safety*.

11 Management of Change
Examples from Practice

Originally published July 2012

Management of change (MOC) is a commonly used technique. Its purpose is to:
- Identify the potential consequences of a change.
- Plan ahead so that counter actions can be taken before a change occurs and continuously as the change progresses.

With respect to operational risks, the process ensures that:
- Hazards are identified and analyzed, and risks are assessed.
- Appropriate avoidance, elimination or control decisions are made so that acceptable risk levels are achieved and maintained throughout the change process.
- New hazards are not knowingly introduced by the change.
- The change does not negatively affect previously resolved hazards.
- The change does not increase the severity potential of an existing hazard.

This process is applied when a site modifies technology, equipment, facilities, work practices and procedures, design specifications, raw materials, organizational or staffing situations, and standards or regulations. An MOC process must consider:
- safety of employees making the changes;
- safety of employees in adjacent work areas;
- safety of employees who will be engaged in operations after changes are made;
- environmental aspects;

- public safety;
- product safety and quality;
- fire protection so as to avoid property damage and business interruption.

OSHA's (1992) Process Safety Management Standard (29 CFR 1910.119) requires that covered operations have an MOC process in place. No other OSHA regulation contains similar requirements, although the agency does address MOC in an information paper (OSHA, 1994). Also, this subject is a requirement to achieve designation in OSHA's Voluntary Protection Programs.

Establishing the Need

Three studies establish that having an MOC system as an element within an operation's risk management system would serve well to reduce serious injury potential. This author reviewed more than 1,700 incident investigation reports, mostly for serious injuries, that support the need for and the benefit of an MOC system. These reports showed that a significantly large share of incidents resulting in serious injury occurs:

- when unusual and nonroutine work is being performed;
- in nonproduction activities;
- in at-plant modification or construction operations (e.g., replacing an 800-lb motor on a platform 15 ft above the floor);
- during shutdowns for repair and maintenance, and startups;
- where sources of high energy are present (electrical, steam, pneumatic, chemical);
- where upsets occur (situations going from normal to abnormal).

Having an effective MOC system will reduce the probability of serious injuries and fatalities occurring in these operational categories.

A 2011 study by Thomas Krause and colleagues produced results that support MOC systems as well. Seven companies participated. Shortcomings in prejob planning, another name for MOC, were found in 29% of incidents that had serious injury or fatality potential. Focusing on reducing that 29%, a noteworthy number, is an appropriate goal. (Data based on personal communication. BST is to publish a paper including these data.)

In personal correspondence, John Rupp of United Auto Workers (UAW) confirmed the continuing history with respect to fatalities occurring in UAW-represented workplaces. According to Rupp, from 1973 through 2007, 42% of fatalities involved skilled-trades workers, who represent about 20% of UAW membership. Rupp also reported that from 2008 through 2011, 47% of fatalities involved skilled-trades workers. These workers are not performing routine production jobs. They often perform unusual and nonroutine work, in-plant modification or construction operations, shutdowns for repair and maintenance, start-ups and near sources of high energy. An MOC (or prejob planning) system would be beneficial for such activities.

Assessing the Need for a Formalized MOC System

Studying an organization's incident experience and that of its industry can produce useful data on the need for a formalized MOC system. Workers' compensation claims experience can be a valuable resource as well.

To develop meaningful and manageable data, an SH&E professional should execute a computer run of an organization's claims experience covering at least 3 years to identify all claims valued at $25,000 or more, paid and reserved. If experience in other organizations is a guide, this run will likely encompass 6% to 8% of the total number of claims and 65% to 80% of the total costs.

Data analysis should identify job titles and incidents that have occurred during changes, and indicate whether a formalized MOC system is needed. Industry experience that may be available through a trade association or similar industry group also should be reviewed. Finding that few incidents resulting in serious injury occurred when changes were being made should not deter an SH&E professional from proposing that the substance of an MOC system be applied to particular changes which present serious injury potential.

Experience Implies Opportunity

To test whether personnel in operations other than chemical sites had recognized the need for and developed MOC systems, the author queried members of ASSE's Management Practice Specialty. The response was overwhelmingly favorable, and the number of example documents received was more than could be practicably used.

Examples received demonstrate that management in various operations has recognized the need for MOC systems. Eight systems selected from this exercise and two previously available are presented as examples in this article. Due to space restrictions, of the 10, one is printed on pp. 210-211.

These select examples show:
- the broad range of harm and damage categories covered;
- similarities in the subjects covered;
- the wide variation in how those subjects are addressed.

These examples reflect real-world applications of MOC in nonchemical operations. They display how such systems are applied in practice.

History Defines Needs and Difficulties of Application

At least 25 years ago, the chemical and process industries recognized the importance of having an MOC process in place as an element within an operational risk management system. That awareness developed because of several major incidents that occurred when changes were taking place.

In 1989, Center for Chemical Process Safety (CCPS) issued *Guidelines for Technical Management of Process Safety,* which included an MOC element. In 1993, Chemical Manufacturers Association published *A Manager's Guide to Implementing and Improving Management of Change Systems.*

In 2008, CCPS issued *Guidelines for Management of Change for Process Safety,* which extends the previous publications. From the preface:

> The concept and need to properly manage change are not new; many companies have implemented management of change (MOC) systems. Yet incidents and near misses attributable to inadequate MOC systems, or to subtle, previously unrecognized sources of change (e.g., organizational changes), continue to occur.
>
> To improve the performance of MOC systems throughout industry, managers need advice on how to better institutionalize MOC systems within their companies and facilities and to adapt such systems to managing non-traditional sources of change. (p. xiii)

Note that incidents and near misses (near hits) attributable to inadequate MOC systems continue to occur. Also, organizational changes are being recognized as a previously unrecognized source from which MOC difficulties could arise. As noted by CCPS (2008), "Management of change is one of the most important elements of a process safety management system" (p. 1).

MOC Requirements in Standards and Guidelines

Several standards and guidelines require or suggest that an MOC process be instituted, including:

•ANSI/AIHA Z10-2005, American National Standard for Occupational Health and Safety Management Systems, which requires that an MOC process be implemented (Section 5.1.2).

•BS OHSAS 18001:2007, Occupational Health and Safety Management Systems Requirements, which states, "For the management of change, the organization shall identify the OH&S hazards, and OH&S risks . . . prior to the introduction of such changes" (Section 4.3.1).

•OSHA comments on change analysis in its Safety and Health Management System eTool: Worksite Analysis.

Anytime something new is brought into the workplace, be it a piece of equipment, different materials, a new process or an entirely new building, new hazards may unintentionally be introduced. Any worksite change should be analyzed thoroughly beforehand because this analysis helps head off problems before they develop.

Provisions that require MOC systems may have different names. For example, Section 7.3.7 of ANSI/ASQ Q9001-2000, Quality Management Systems: Requirements, is titled "Control of design and development changes." It states:

> Design and development changes shall be identified and records maintained. The changes shall be reviewed, verified and validated, as appropriate, and approved before implementation. The review of design and development changes shall include evaluation of the effect of the changes on constituent parts and product already delivered. Records of the results of the review of changes and any necessary actions shall be maintained.

The MOC Process

As with all management systems, an administrative procedure must be written to communicate what the MOC system encompasses and how it should operate. The system should be designed to be compatible with the organization's and industry's inherent risks; management systems in place; organizational structure; dominant culture; and expected workforce participation.

Although brevity is the goal, several subjects should be considered for inclusion in an MOC procedure:

1) Define the need for and the purpose of an MOC system.
2) Establish accountability levels.
3) Specify criteria that will trigger formal change requests.
4) Specify how personnel will submit change requests and what form will be used.
5) Outline criteria for request reviews, as well as responsibilities for those reviews.
6) Indicate that the MOC system encompasses:
•risks to those performing the work and other affected employees;
•possible property damage and business interruption;
•possible environmental damage;
•product safety and quality;
•procedures to accomplish the change;
•results evaluation.
7) Establish that minute-by-minute control must be maintained to achieve acceptable risk levels, and that risk assessments will be made as often as needed while work progresses; this will involve giving instruction on needed action if unanticipated risks of concern are encountered.

8) Identify who will accept or decline a change request, including an MOC approval form.

9) Outline a method to determine the actions necessary because of the effect of changes (e.g., providing more employee training; revising standard operating procedures and drawings; updating emergency plans).

10) Indicate that work will receive a final review before startup of operations, and identify the titles of those who will conduct the review.

Responsibility Levels

In drafting an MOC system, responsibility levels must be defined and must align with an entity's organizational structure. This is a critical step in developing an MOC system. If even minor process changes are considered critical with respect to employee injury and illness potential, possible environmental contamination, and product quality and safety, then the levels of responsibility are often many. Some systems used as examples in this article clearly establish responsibility levels, while others do not.

Examples of responsibility levels, as outlined in an organization where inherent hazards require close control, are provided as reference points.

•**Initiator:** The initiator owns the change and is responsible for initiating the change request form. Based on complexities of the changes, these responsibilities may be reassigned at any time. The initiator fully describes and justifies changes; ensures that all appropriate departments have assessed changes; manages execution of the change request; and ensures that the changes are implemented properly.

•**Department supervisor:** The department supervisor assigns qualified personnel to initiate change requests. The change control process is critical to employee safety, as is avoiding environmental contamination and ensuring product quality. This supervisor ensures that the change request is feasible and adequately presented for review.

•**Document reviewers:** Document reviewers assess and approve change request forms. These activities include reviewing the document for accuracy and adequacy with respect to proposed changes.

•**Approvers:** Department managers should select preapprovers with expertise related to the nature of the proposed change. Based on knowledge and expertise, each reviewer will evaluate and assess the effect of the proposed change on existing processes in his/her area of expertise. Reviewers also must review and approve the change request form and the implementation plan to evaluate the change and ensure that the steps for implementation are appropriate. This is the final review before the proposed change is implemented.

•**Postimplementation approvers:** Department managers should select postimplementation approvers who should ensure that the change has been appropriately implemented as indicated when approval for the requested change was given. This process also ensures that only the changes shown on the change request form have been implemented.

The MOC Process: Activities to Consider

An organization's hazard and risk complexities and the desire to establish an adequate yet not complex MOC system should be considered when identifying activities that will activate that system. Activity categories may include the following:
- Nonroutine and unusual work is to be performed.
- Work exposes workers to sources of high energy.
- Maintenance operations for which prejob planning and safety reviews would be beneficial because of inherent hazards.
- Substantial equipment replacement work.
- Introduction of new or modified technology, including changes to programmable logic controllers.
- Modifications are made in equipment, facilities or processes.
- New or revised work practices or procedures are introduced.
- Design specifications or standards are changed.
- Different raw materials will be used.
- Safety and health devices and equipment will be modified.
- The site's organizational structure changes significantly.
- Changes in staffing levels that may affect operational risks.
- Staffing changes require a review of skill levels.
- The site changes how it uses contractors.

MOC Request Form

An MOC request form is needed, and its content should align with an organization's structure and in-place management systems (e.g., capital request procedures, work orders, purchasing procedures). Creating a digital form allows flexibility for descriptive data and comments. The form should include:
- name of person initiating the request;
- date of request;
- department or section or area;
- equipment, facility or processes affected;

- brief description of proposed change and what will be accomplished;
- potential performance and SH&E considerations;
- titles and names of personnel who will review the change;
- effect on standard operating procedures, maintenance, training and similar functions;
- space for reviewers to document special conditions or requirements;
- approvals and authorizations;
- routing indicators or provisions for copies to be sent to personnel responsible for training and updating operating procedures, drawings and similar documents.

Implementing the MOC Process

Senior management and safety professionals must appreciate the magnitude of the task of initiating and implementing an MOC system, and should expect pushback. Common obstacles include egos, territorial prerogatives, the current power structure and normal resistance to change; remember, those affected may have had little experience with the administrative systems being proposed. Although MOC systems have been required in the chemical industries for many years, the literature reports that their application has experienced difficulties. According to CCPS (2008):

> Even though the concept and benefits of managing change are not new, the maturation of MOC programs within industries has been slow, and many companies still struggle with implementing effective MOC systems. This is partly due to the significant levels of resources and management commitment that are required to implement and improve such systems. MOC may represent the biggest challenge to culture change that a company faces. (p. 10)
>
> Developing an effective MOC system may require evolution in a company's culture; it also demands significant commitment from line management, departmental support organizations, and employees. (p. 11)

Management commitment, evidenced by providing adequate resources and the leadership required to achieve the necessary culture change, must be emphasized. Stated or written management commitment that is not followed by providing the necessary resources is not management commitment. Because of extensive procedural revisions necessary when initiating an MOC system, culture change methods should be applied. Subjects to consider when implementing the system include the following:

- Management commitment and leadership must be obtained and demonstrated. That means providing personal direction and involvement in initiating procedures; providing adequate resources; and making appropriate decisions with respect to safety when disagreement arises about the change review process.

- Keep procedures as simple as practicable. An applied, less-complicated system achieves better results than an unused complex system.
- Obtain widespread acceptance and commitment. Inform all affected employees before initiating the MOC system, solicit their input, and respect their perspectives and concerns.
- Recognize the need for and provide necessary training.
- Field-test a system before implementing it. Debugging will produce long-term returns.
- After refining the system through a field test, select a job or an activity that would benefit—both productivity/efficiency and safety—from an MOC system, and emphasize those benefits to build favorable interest. Testing the system in a select activity demonstrates its value, makes it credible and creates demand for additional applications.
- Monitor system progress and performance via periodic audits, and informally ask employees for their perspectives.

Managing Organizational Change

In some examples posted, procedures require those involved to assess the significance of organizational changes. These provisions exist because organizational and personnel changes can negatively affect an operational risk management system.

Of the considerable literature on the subject, Managing the Health and Safety Impacts of Organizational Change (CSChE, 2004) is cited because it fits closely with the intent of some examples provided. Types of organizational and personnel changes that can negatively affect operations risk management are:

- reorganizing or reengineering;
- workforce downsizing;
- attrition and workforce aging;
- outsourcing of critical services;
- changes affecting the competence or performance of contractors that provide critical services (e.g., equipment design, process control software, hazard and risk assessment);
- loss of skills, knowledge or attitudes as a result of the cited changes.

According to CSChE (2004), such changes are not as well-addressed in applicable guidelines as changes in equipment, tools, work methods and processes.

More emphasis must be given to the effect of organizational changes on operational risk management because incident reports on some serious injuries and fatalities indicate that staffing reductions were a significant contributing factor to unacceptable risk situations such as inadequate maintenance; inadequate competency; workers being stressed beyond their mental and physical capabilities (e.g., two persons doing the work for which three had been previously assigned; a person working alone in a high-hazard situation for which the standard operating procedure calls for a work buddy).

Risk Assessments

Some of the examples posted require risk assessments at several stages of the change activity. The intent is to achieve and maintain acceptable risk levels throughout the work. Thus, risk assessments should be conducted as often as needed as changes occur—and particularly when unexpected situations arise. SH&E professionals who become skilled risk assessors can offer a significant value-added consultancy.

Risk assessment is the core of ANSI/ASSE Z590.3, Prevention Through Design: Guidelines for Addressing Occupational Hazards and Risks in Design and Redesign Processes. The standard's content is applicable to MOC whether the contemplated change involves new designs or redesign of existing operations. Of particular interest are sections on supplier relationships, safety design reviews, hazard analysis and risk assessment processes and techniques, and the hierarchy of controls.

Risk Assessment Matrixes

Z590.3 recommends use of a risk assessment matrix, and stresses that all involved agree on the definitions of terms used in the matrix. An addendum in Z590.3 provides several sample matrixes. The example presented in Figure 1 was preferred by operating employees involved in the risk assessment process. They indicated that first establishing a mental relationship between numbers such as 6 and 12 helped them more readily understand the relation between terms such as *moderate risk* and *serious risk*.

The Significance of Training

CCPS (2008) emphasizes the significance of training in achieving a successful MOC system:

> Training for all personnel is critical. Many systems failed or encountered severe problems because personnel did not understand why the system was necessary, how it worked and what their role was in the implementation. (p. 58)

The culture change necessary to implement a successful MOC system is impossible without a training program that helps supervisors and workers understand the

Severity levels and values	Occurrence probabilities and values				
	Unlikely (1)	Seldom (2)	Occasional (3)	Likely (4)	Frequent (5)
Catastrophic (5)	5	10	15	20	25
Critical (4)	4	8	12	16	20
Marginal (3)	3	6	9	12	15
Negligible (2)	2	4	6	8	10
Insignificant (1)	1	2	3	4	5

Incident or Exposure Severity Descriptions

Catastrophic: One or more fatalities, total system loss and major business down time, environmental release with lasting effect on others with respect to health, property damage or business interruption.

Critical: Disabling injury or illness, major property damage and business down time, environmental release with temporary impact on others with respect to health, property damage or business interruption.

Marginal: Medical treatment or restricted work, minor subsystem loss or property damage, environmental release triggering external reporting requirements.

Negligible: First-aid or minor medical treatment only, nonserious equipment or facility damage, environmental release requiring routine clean-up without reporting.

Insignificant: Inconsequential with respect to injuries or illnesses, system loss or downtime, or environmental release.

Incident or Exposure Probability Descriptions

Unlikely: Improbable, unrealistically perceivable.
Seldom: Could occur but hardly ever.
Occasional: Could occur intermittently.
Likely: Probably will occur several times.
Frequent: Likely to occur repeatedly.

Risk Levels: Combining the severity and occurrence probability values yields a risk score in the matrix. The risks and the action levels are categorized below.

Risk Categories, Scoring and Action Levels

Category	Risk score	Action level
Low risk	1 to 5	Remedial action discretionary.
Moderate risk	6 to 9	Remedial action to be taken at appropriate time.
Serious risk	10 to14	Remedial action to be given high priority.
High risk	15 or greater	Immediate action necessary. Operation not permissible except in an unusual circumstance or as a closely monitored and limited exception with approval of the person having authority to accept the risk.

Figure 1: Risk Assessment Matrix
Note. Numbers were intuitively derived. They are qualitative, not quantitative. They have meaning only in relation to each other.

concepts to be applied. Where the MOC system applies to many risk categories (occupational, public, environmental, fire protection and business interruption, product quality and safety), training must be extensive.

Documentation

An operation must maintain a history of operational changes. All modifications must be recorded in drawings, prints and appropriate files; they become the historical records that would be reviewed when future changes are made.

Comments that "changes made were not recorded in drawings, prints and records" are too common in reports on incidents with serious consequences. Examples of unrecorded changes include the following:
- The system was rewired.
- A blank was put in the line.
- Control instruments were disconnected.
- Relief valves of lesser capacity had been installed.
- Sewer line sensors to detect hazardous waste were removed.

On the MOC Examples

To demonstrate the substance and variety of actual MOC systems, few changes were made in the examples. In some cases, terms used are not readily understandable. However, these terms are likely understood within the organization that developed the system, so they are presented as-is to emphasize that the terminology included in an MOC procedure must reflect the language commonly used within an organization and must be understood by all involved in an MOC initiative.

These examples vary greatly in content and purpose. Some are one page; others take several pages to cover the complexity of procedures and exposures. Some procedures have introductory statements on policy and procedure, others do not. Nevertheless, these examples show that an MOC system need not meet a theoretical ideal to provide value. These examples are intended as references; none should be adopted as is. An MOC system should reflect an organization's particular needs and its culture.

Example 1: Producing Mechanical Components

This prejob planning and safety analysis system (Figure 2) is a one-page outline developed because of adverse occupational injury experience in work that was often unusual or one-of-a-kind, or that required extensive, complex maintenance.

It is relatively simple in relation to other examples posted, yet it was successfully applied for its purposes. In this case, safety professionals:

- prepared the data necessary to convince management and shop floor personnel to try the proposed system;
- reported that training was highly significant in achieving success;
- emphasized that work situations discussed in training were real to that organization;
- addressed productivity/efficiency and risk control benefits in their proposal and during training.

For whoever initiates an MOC system, the following procedure will be of interest.

> At a location where the serious injury experience was considered excessive for nonroutine work, safety professionals decided that something had to be done about it. As they prepared a course of action and talked it up at all personnel levels, from top management down to the worker level, they encountered the usual negatives and push-back (e.g., it would be time consuming, workers would never buy into the program, supervisors would resist the change). The safety professionals considered the negatives as normal expressions of resistance to change.
>
> Their program consisted of, in effect, indoctrinating management and the workforce in the benefits to be obtained by doing pre-reviews of jobs so that the work could be done effectively and efficiently while at the same time controlling the risks.
>
> Eventually, management and the line workers agreed that classroom training sessions could be held. Later, the safety professionals said that the classroom training sessions and follow-up training were vital to their success.
>
> At the beginning of each of those sessions, a management representative introduced the subject of prejob planning and safety analysis and discussed the reasons why the new procedure was being adopted. Statistics on accident experience prepared by safety professionals were a part of that introduction. Then, safety professionals led a discussion of the outline shown in MOC Example 1. It set forth the fundamentals of the prejob review system being proposed. After discussion of those procedures, attendees were divided into groups to plan real-world scheduled maintenance jobs that were described in scenarios that had been previously prepared.
>
> At this location, supervisors took to the pre-job planning and safety analysis system when they recognized that the system made their jobs easier, improved productivity/efficiency, and reduced the risks. And they took ownership of the system. As one of the safety professionals said, "Our supervisors and workers have become real believers in the system." And a culture change had been achieved.

Note the requirements under the caption "Upon Job Completion." The detail of the requirements reflects particular incidents with adverse results that occurred over several years. Every MOC system should include similar procedures to be followed before work can be considered completed.

Example 2: Specialty Construction Contractor

This two-page field work review and hazard analysis system was provided by a safety professional employed by a specialty construction contractor that has several crews

Alpha Corporation

Pre-Job Planning and Safety Analysis Outline

1) Review the work to be done. Consider both productivity and safety:
   a) Break the job down into manageable tasks.
   b) How is each task to be done?
   c) In what order are tasks to be done?
   d) What equipment or materials are needed?
   e) Are any particular skills required?
2) Clearly assign responsibilities.
3) Who is to perform the pre-use of equipment tests?
4) Will the work require: a hot work permit; a confined entry permit, lockout/tagout (of what equipment or machinery), other?
5) Will it be necessary to barricade for clear work zones?
6) Will aerial lifts be required?
7) What personal protective equipment will be needed?
8) Will fall protection be required?
9) What are the hazards in each task? Consider:

   - Access
   - Chemicals
   - Conveyors
   - Dropping tools
   - Dusts
   - Electricity
   - Elevated loads
   - Explosion
   - Fall Hazards
   - Fire
   - Forklift trucks
   - Hot objects
   - Machine guarding
   - Moving equipment
   - Noise
   - Pressure
   - Sharp objects
   - Steam
   - Stored energy
   - Twisting, bending
   - Vibration
   - Weather
   - Weight of objects
   - Welding
   - Work at depths
   - Work at heights
   - Worker position
   - Worker posture

10) Of the hazards identified, do any present severe risk of injury?

11) Develop hazard control measures, applying the Safety Decision Hierarchy.
   • Eliminate hazards and risks through system and work methods design and redesign
   • Reduce risks by substituting less hazardous methods or materials
   • Incorporate safety devices (fixed guards, interlocks)
   • Provide warning systems
   • Apply administrative controls (work methods, training, etc.)
   • Provide personal protective equipment
12) Is any special contingency planning necessary (people, procedures)?
13) What communication devices will be needed (two-way, hand signals)?
14) Review and test the communication system to notify the emergency team (phone number, responsibilities).
15) What are the workers to do if the work doesn't go as planned?
16) Considering all of the foregoing, are the risks acceptable? If not, what action should be taken?

Upon Job Completion
17) Account for all personnel
18) Replace guards
19) Remove safety locks
20) Restore energy as appropriate
21) Remove barriers/devices to secure area
22) Account for tools
23) Turn in permits
24) Clean the area
25) Communicate to others affected that the job is done
26) Document all modifications to prints and appropriate files

Figure 2: MOC Example 1

active in various places at the same time. Note that the names of employees on a job must be documented as having been briefed on the work to be performed. The checklist included in this example pertains to occupational, public and environmental risks.

When asked what drove development of the change procedure, the safety professional said the firm had learned from costly experience. According to this professional, the procedures required by the change system are now embedded in the company's operations and are believed to have resulted in greater efficiency. In addition, fewer costly incidents have occurred.

This example has a direct relation to the purposes of construction/demolition standard ANSI/ASSE 10.1, Pre-Project and Pre-Task Safety and Health Planning. This standard is an excellent resource for contractors and companies that establish requirements for on-site contractors. Note the distinctions: preproject planning and pretask planning.

Example 3: Serious Injury Experience

This pretask analysis form emphasizes obtaining required permits and ensuring that supervisors brief employees on the order of activities and about the risks to be encountered. Employees must sign the form to confirm this briefing; it is the only example that requires employee signatures.

Example 4: Management Policy and Procedure

This basic guidance paper is condensed to three pages. It is a composite of several MOC policies and procedures issued by organizations in which operations were not highly complex. Reference is made to an MOC champion. Someone must be responsible for the change and manage it through to an appropriate conclusion.

Example 5: Specifically Defined Prescreening Questionnaire

In three pages, this system commences with an interesting prescreening questionnaire. If the answer to all questions is "no," the formal MOC checklist and approval form need not be completed. With respect to MOC systems, it's often asked, "To what work does the system apply?" This organization developed a way to answer that question for its operations.

Example 6: High-Risk Multiproduct Manufacturing

This MOC policy and procedure reflect the organization's high-hazard levels. Captions in this four-page example are: safety; ergonomics; occupational health; radiation control; security/property loss prevention; clean air regulations; spill prevention and community planning; clean water regulations; solid and hazard waste regulations; environmental, safety and health management systems; and an action item tracking instrument.

According to the procedure paper announcing the system:

> If a significant change occurs with respect to key safety and health or environmental personnel, the matter will be reviewed by the S&H manager and the environmental manager and a joint report including a risk assessment and their recommendations will be submitted to location management.

Example 7: A Food Company

This four-page policy includes product safety and quality as subjects to be considered. The safety director reports that discipline in the application of this MOC system is rigid, which reflects management's determination to avoid damaging incidents and product variations. Provisions for prestart-up and postmodification are extensive. Risks are assessed after changes are made and before start-up.

Example 8: Conglomerate

Iota Corp. has a five-page MOC procedure outlined in four sections. Section I requires completion of a change request and tracking system requirements form in which the change is described, a tracking number is assigned and approval levels are established. Approval levels are numerable, including headquarters in some instances.

Section II outlines a change review and approval procedure that is extensive with respect to occupational safety, health and environmental concerns. Section III is a preimplementation action summary form that lists subjects for which actions are necessary before change can begin, and it identifies those responsible for those actions. Section IV lists 11 points in a postcompletion form.

Example 9: Extensive System for a Particular Operation

This seven-page example is valuable because of its structure and content. It is somewhat different compared to the other examples. It:

•handles requirements for technical changes and organizational changes separately and extensively;

•stresses organizational changes for which risk assessments are required;

•outlines in detail technical changes to which the standard applies;

•lists risk assessment as a separate item that pertains to all operations;

•includes a lengthy discussion of general considerations, and outlines and thoroughly discusses requirements for a six-point MOC standard: management process; capability; change identification; risk management; change plan; and documentation.

This is a concept and procedural paper. It does not include the forms used to implement the various procedures.

Example 10: International Multioperational Entity

Application of this system, titled Management of Change Policy for Safety and Environmental Risks, extends the activities of SH&E professionals beyond that of any other example. It covers 10 pages, although the bulletin issued by the safety entity within the organization is much longer.

Only two of five exhibits were made available because of proprietary reasons. Unique aspects include the following:

•Due diligence is included in a list of definitions. SH&E professionals are to assess acquisitions and similar transactions.

•Global franchise management board members are listed under responsibilities. They are to ensure compliance with the standard.

•A preliminary SH&E assessment questionnaire shall be initiated during the project planning stage.

•A section titled Evaluating Change (Risk Assessment Guidelines) includes these subjects, which may not be included in other examples, at least not as extensively:

a) new process product and development;

b) capital/noncapital project;

c) external manufacturing;

d) business acquisitions;

e) significant downsizing/hiring.

•Conducting risk analyses is a major section.

This example is noteworthy because of its breadth. Interestingly, the system was issued by the safety and industrial hygiene unit, which implies management support for superior operational risk management.

Conclusion

This article has provided a primer that SH&E professionals can use to craft an organization-specific MOC/prejob planning system. Safety professionals should consider whether their employers could benefit from having such a system in place. Having a system that prestudies changes because of their inherent hazards and risks and their potential effect on safety, productivity and environmental controls is good risk management. ∎

References

American Society for Quality (ASQ). (2000). Quality management systems: Requirements (ANSI/ASQ Q9001-2000). Milwaukee, WI: Author.

ANSI/ASSE. (2011). Construction and demolition operations: Preproject and pretask safety and health planning (ANSI/ASSE A10.1-2011). Des Plaines, IL: Author.

ANSI/ASSE. (2011). Prevention through design: Guidelines for addressing occupational hazards and risks in design and redesign processes (ANSI/ASSE Z590.3-2011). Des Plaines, IL: Author.

Canadian Society of Chemical Engineering (CSChE). (2004). Managing the health and safety impacts of organizational changes. Ottawa, Canada: Author. Retrieved from www.cheminst.ca/index.php?ci_id=3210&la_id=1.

Center for Chemical Process Safety (CCPS). (2008). *Guidelines for management of change for process safety*. Hoboken, NJ: American Institute of Chemical Engineers, Author.

Chemical Manufacturers Association. (1983). *A manager's guide to implementing and improving management of change systems*. Washington, DC: Author.

NSC. (2000). *Aviation ground operations safety handbook* (5th ed.). Itasca, IL: Author.

OSHA. Safety and health management system eTool: Worksite analysis. Washington, DC: U.S. Department of Labor, Author. Retrieved from www.osha.gov/SLTC/etools/safety health/comp2.html.

OSHA. (1992). Process safety management of highly hazardous chemicals (29 CFR 1910.119). Washington, DC: U.S. Department of Labor, Author.

OSHA. (1994). Process safety management guideline for compliance (OSHA Bulletin 3133). Washington, DC: U.S. Department of Labor, Author. Retrieved from www.osha.gov/Publications/osha3133.pdf.

Stephens, R.A. (2004). *System safety for the 21st century*. Hoboken, NJ: John Wiley & Sons.

12 PREVENTING SERIOUS INJURIES AND FATALITIES
TIME FOR A SOCIOTECHNICAL MODEL FOR AN OPERATIONAL RISK

ORIGINALLY PUBLISHED MAY 2013

Results of recent attempts to reduce serious injuries and fatalities cannot be considered stellar. In 2007, a national forum on Fatality Prevention in the Workplace was sponsored by Indiana University of Pennsylvania in cooperation with the Alcoa Foundation. Many speakers suggested tweaking elements in existing occupational risk management systems.

At about the same time, ORC Worldwide (now Mercer HSE Networks), an organization whose members represent about 120 of the Fortune 500 companies, conducted a study to identify the characteristics of serious injuries and fatalities. The intent of the study, which was partially achieved, was to provide member companies information on how to improve reduction efforts.

In announcing the Alcoa Foundation grant to support the fatality prevention forum, Lon Ferguson (2007) said:

> Reliance on traditional approaches to fatality prevention has not always proven effective. This fact has been demonstrated by many companies, including some thought of as top performers in safety and health, as they continue to experience fatalities while at the same time achieving benchmark performance in reducing less-serious injuries and illnesses.

Ferguson's statement still applies, particularly the idea that "reliance on traditional approaches to fatality prevention has not always proven effective." Companies with outstanding records showing reductions in less-serious injuries may not have had similar reductions for serious injuries and fatalities. At Mercer HSE Networks, about 40 companies are involved in a study to determine what can be done to reduce occupational fatalities. Such studies are important, but major innovations in safety management systems are needed as well. Tweaking systems in place will not achieve the substantial improvements desired.

Scope of This Article

Serious injuries and fatalities are treated as one subject in this article for several reasons. Many serious injuries could have been fatalities under slightly different circumstances. Thus, data on serious injuries (and near-misses) should be analyzed because the results may provide valuable information on actions to be taken to prevent fatalities and other serious injuries. In addition, causal factors for serious injuries and the actions necessary to prevent them are identical to those for fatalities. Many organizations do not have fatalities, but they may have serious injuries. As will be discussed, data derived from analyses of serious injuries may be influential in achieving attention to incidents that have fatality potential.

A Statistical Review Serious Injury Trending

A 2005 National Council on Compensation Insurance (NCCI) research brief states that "there has been a larger decline in the frequency of smaller lost-time claims than in the frequency of larger lost-time claims." As Table 1 shows, the reduction in cases valued from $10,000 to $50,000 is about one third of that for cases valued at less than $2,000. For cases valued above $50,000, the reduction is about one fifth of that for the less-costly injuries. Thus, costly claims—those for serious injuries and fatalities—loom larger within the spectrum of all claims reported.

Value of claim	Reduction
Less than $2,000	34%
$2,000 to $10,000	21%
$10,000 to $50,000	11%
More than $50,000	7%

Table 1: Claims Frequency Trends
Note. Data from 1999 and 2003, expressed in 2003 hard dollars. Adapted from "State of the Line," by D. Mealy, 2005, National Council on Compensation Insurance News Bulletin.

In 2011, NCCI (Davis & Bar-Chaim) reported that from 2005 through 2009, "after accounting for wage and medical cost inflation, claims below $50,000 exhibited a greater rate of decline than those above $50,000" (Table 2). These data are in concert, generally, with the trending shown for the years 1999 through 2003. Reductions in less-serious injuries are substantially more than those for more serious injuries. In its 2009 State of the Workers' Compensation Line Report (Mealy, 2009), NCCI reported that injury frequency had declined consistently for all injury types except for permanent total disabilities.

In 2011, NCCI also noted that "workers' compensation claim frequency for lost-time claims has increased 3% in 2010. This represents the first increase since 1997 and only the third time that frequency has increased in the last 20 years." This upward claim frequency trend is relative to the increase in fatalities for 2010 (Table 3).

Claim value	Reduction
Less than $2,000	25%
$2,000 - $10,000	22%
$10,000 - $50,000	20%
$50,000 - $250,000	14%
More than $250,000	9%

Table 2: Serious Injury Claims Decline Lags
Note. Adapted from "Workers' Compensation Claim Frequency," by J. Davis and Y Bar-Chaim, 2011, National Council on Compensation Insurance.

Year	Fatalities	Fatality rate
1971	13,700	17
1981	12,500	13
1991	9,800	8
2001	5,900	4.3
2002	5,524	4.0
2003	5,559	4.0
2004	5,703	4.1
2005	5,702	4.0
2006	5,703	3.9
2007	5,488	3.7
2008	5,214	3.7
2009	4,551	3.5
2010	4,690	3.6
2011	4,608	3.5*

Table 3: All Fatalities, All Occupations: 1971 to 2011
Note. *Data for 2011 are preliminary. Adapted from Accident Facts, by National Safety Council, 1995, Itasca, IL: Author; and "Census of Fatal Occupational Injuries, 1996-2011," by Bureau of Labor Statistics, Washington, DC: Author, U.S. Department of Labor.

Fatality Trending

Data presented in Table 3 are based on excerpts from Accident Facts (NSC, 1995) and the Bureau of Labor Statistics (BLS) Census of Fatal Occupational Injuries, 1996-2011. In both datasets, the fatality rate is the number of fatalities per 100,000 workers.

Data for 2011 are preliminary. BLS expects to issue a final report for the 2011 year in April 2013 (as this issue went to press). For previous years, the average increase in the number of fatalities in the final report was 166. Add that number to 4,608 and the total is 4,775—an increase over 2010 even though fewer people were working with fewer hours of exposure.

Reductions in the number of fatalities and fatality rates are huge and commendable, and they indicate growth in management enlightenment, technology improvements, and an extended application of hazard identification, risk assessment, and avoidance and reduction techniques.

The data in Table 3 indicate that the fatality record has plateaued in recent years. Fatality rates over the past 6 years range from 3.9 to 3.5 with an average of 3.7. Some might say that the low-hanging fruit has been picked and that achieving further substantial reductions will require exceptional efforts.

Developing Fatality Data for Individual Industry Categories

All of the preceding data is macro. It pertains to all occupations. Further studies of industry categories are needed to examine trending of fatalities and fatality rates and the types of activities in which fatalities occur. This is so that innovations aimed at reducing serious injuries and fatalities should pertain specifically to operational needs. While generalities can be suggested with respect to the content and order of elements in an occupational risk management system, emphasis in the application of those elements should result from studies that determine where the largest opportunities lie and where the emphasis should be.

Building Interest in Serious Injury and Fatality Prevention

For this discussion, data for the manufacturing industry were selected to illustrate that annual fatality rates are in a narrow statistical range and that the statistical probability of an organization having a fatality is low. Methods to achieve an interest in the subject are also discussed.

Consider the data presented in Table 4. (Select any industry and comparable data can be produced.) Fatality rates fall within a very narrow range. For the 6 years shown, the average fatality rate is 2.27. This author projected data contained in a once-in-5-years report issued by the U.S. Census Bureau for 2007 and estimated that there were about 300,000 manufacturing locations in the U.S. in 2010. BLS (2011) reports that manufacturing had 11,575,000 employees in that year. A small percentage of manufacturers reported the 320 fatalities that occurred in 2010. Many have never had a fatality.

Year	Fatalities	Fatality rate per 100,000 employees
2006	447	2.1
2007	392	2.4
2008	389	2.5
2009	304	2.2
2010	320	2.2
2011	322	2.2

Table 4: Manufacturing
Note. Adapted from "Manufacturing Employment," by Bureau of Labor Statistics. Washington, DC: Author.

To focus needed attention on serious injury and fatality prevention in organizations that have had no or very few fatalities over a period of many years, SH&E professionals should focus on the potential for serious injuries and fatalities based on:
- serious injuries that occurred;
- less-than-serious injuries that could have been serious in other circumstances;
- selected near-misses that had serious injury potential.

As noted, reducing the number of serious injuries and fatalities will require major innovations in occupational risk management systems.

Innovations to Be Considered

Bringing the necessary attention to serious injury and fatality prevention will require enormous culture changes as well as recognition of how deeply some deterring premises are embedded in many companies. Following are several innovations to consider, and other safety professionals may want to add to the list.
- The premise that OSHA-related incidence rates are accurate measures of serious injury and fatality potential must be dislodged.
- The belief that unsafe acts of workers are the principal causes of occupational incidents must be uprooted and dislodged.

- The broadly held assumption that reducing the frequency of less-than-serious injuries will result in an equivalent reduction in serious injuries must be dislodged.
- Risk assessments must be recognized and established as the core of an operational risk management system.
- Prevention through design concepts must be instituted as an element within an operational risk management system.
- Businesses must understand the ongoing transition concerning the prevention of human error, which directs prevention efforts to the design of the work system and work methods.
- Management of change/prejob planning must be a separate and emphasized element within an operational risk management system.
- Incident investigations must be improved so that shortcomings in management systems related to serious injury and fatality potential can be identified and addressed.
- Internally published operational risk management systems must be revised in relation to the foregoing.

Although this article focuses on serious injury and fatality prevention, improving on or instituting these innovations will help reduce injuries at all severity levels.

Achieving a Culture Change

It will take a major educational undertaking to convince management, and subsequently all personnel, that achieving low OSHA incident rates does not indicate that controls are adequate with respect to serious injury and fatality potentials. For more than 40 years, low OSHA incident rates have been overemphasized, resulting in competition within companies and among companies within an industry group. When achieving low OSHA incidence rates is deeply embedded within an entity's culture, uprooting and dislodging it will be a challenging, long-term effort. A culture change is not a one-time activity. It is a long journey that must engage all members of an organization.

In the culture change process, SH&E professionals must make the case that priority attention be given to recognizing and avoiding hazardous situations with serious injury potential. This approach must be tailored to the given entity's needs and opportunities. For example, consider these three potential courses of action.

1) Collect all incident investigation reports for a 3-year period, then select those that describe situations for which, under slightly different circumstances, the results could have been a more serious injury or fatality. This process could lead to analyses

of operations in which the incidents occurred and advance the idea that serious injury potential needs special consideration.

2) Request a report of all workers' compensation claims valued at $25,000 or more for 3 years. Why this level? In the author's experience, a $25,000 cut-off level returns 6% to 8% of the total number of claims and 60% to 80% of total claims values. Of course, there have been outliers. For example, for a manufacturing company that also had a mining operation, 25% of cases valued at $25,000 or more represented 75% of total claims value. In another organization, 90% of total claims value came from 5% of the claims valued at $25,000 or more. In each case, valuable data were produced. Even for very large companies, the report output has been manageable.

3) Engage employees in an information gathering system that continuously reports on hazardous situations with serious injury potential. The system should include near-misses that could have resulted in serious consequences under slightly different circumstances. To succeed, the company must understand that employees who are encouraged to provide input are recognized as valuable resources because of their extensive knowledge about how work is performed. Also, they must be respected for their knowledge and skills. Feedback on employee input is a must.

The data collected should be mined for information to support a proposal that serious injury and fatality potential must be a focus. Specifically, job titles, units or departments that are prominent in the data should be reviewed, as should the types of operations in which the injuries occurred. For example, in several companies, 60% to 80% of injuries valued at $25,000 or more involved employees that were not making product.

SH&E professionals must also identify other methods to produce meaningful and convincing data relating to the inherent risks in an organization and its culture. This presents an opportunity for creativity.

Countering the Premise That Unsafe Acts Are the Principal Cause of Occupational Incidents

Manuele (2011) suggests that two myths related to the work of H.W. Heinrich need to be dislodged from the practice of safety. One of those myths is the premise that workers' unsafe acts are the principal cause of occupational incidents. Manuele addresses topics such as moving preventive efforts from a focus on employee to a focus on the work system; the design of the work system and the work methods; the complexity of causation; and human error occurring at organizational levels above the worker.

Response in support of the article was good. The feedback received indicates how deeply Heinrich's premise that 88% of accidents are caused by workers' unsafe acts is embedded in the organizations that safety practitioners advise. To more effectively prevent serious injuries and fatalities, this myth clearly must be dislodged from the practice of safety.

Reducing Injury Frequency Will Not Equivalently Reduce Severity

Heinrich's premises with respect to what had become broadly known as his 300-29-1 ratios varied in the first three editions of his book *Industrial Accident Prevention*. The following statement appeared in the third and fourth editions.

> Analysis proves that for every mishap resulting in an injury there are many other similar accidents that cause no injuries whatever. From data now available concerning the frequency of potential-injury accidents, it is estimated that in a unit group of 330 accidents *of the same kind and involving the same person* [emphasis added], 300 result in no injuries, 29 in minor injuries and 1 in a major or lost-time injury.

On its face, this statement cannot be substantiated. Heinrich also wrote that "in the largest injury group—the minor injuries—lies the most valuable clues to accident causes." That became the premise from which educators taught and many safety practitioners came to believe that reducing accident frequency will achieve equivalent reduction in injury severity. This myth is also deeply embedded in the minds of some safety practitioners and the management personnel whom they advise. It must also be dislodged. Statistical data presented in this article refute the premise. Based on thorough research through all four editions of Heinrich's book, the author has concluded that the 300-29-1 ratios are not supportable.

On Risk Assessments

Risk assessments should be established as the core of an operational risk management system as a separately identified element following the first element that would be comparable to management leadership, commitment, demonstrated involvement and accountability.

Europeans have long advocated risk assessments as a core value in injury and illness prevention. Other activity around the globe also promotes risk assessments. For example:

1) Guidance on the Principles of Safe Design for Work issued in 2006 by the Australian Safety and Compensation Council, an entity of the Australian government,

includes a risk management process and promotes integrating risk management into the design process.

2) Requirements for risk assessments are more explicit in the 2007 revision of BS OHSAS 18001:2007, Occupational Health and Safety Management Systems—Requirements. Commonly known as 18001, this British Standards Institution publication states, "The organization shall establish, implement and maintain a procedure(s) for the ongoing hazard identification, risk assessment and determination of necessary controls."

3) In 2008, the U.K. Health and Safety Executive issued "Five Steps to Risk Assessment." By law, all employers in the U.K. must conduct risk assessments.

4) ANSI B11.0: Safety of Machinery—General Safety Requirements and Risk Assessments applies to a broad range of machinery. Note that *risk assessments* is included in the title. It describes procedures for identifying hazards, assessing and reducing risks to an acceptable level over the life cycle of machinery.

5) In March 2011, DOT's Pipeline and Hazardous Materials Safety Administration proposed modification of HazMat regulations to require that risk assessments be made of loading and unloading operations.

6) In August 2008, the European Union (EU) launched a 2-year campaign focusing on risk assessment. EU says this about the campaign:

> Risk assessment is the cornerstone of the European approach to prevent occupational accidents and ill health. If the risk assessment process—the start of the health and safety management approach—is not done well or not at all, the appropriate preventive measures are unlikely to be identified or put in place.

It is highly significant that the EU declared that "risk assessment is the cornerstone of the European approach to prevent occupational accidents and ill health." That statement is foundational and should be supported by safety professionals. Johnson (1980) expressed a companion view: "Hazard identification is the most important safety process in that, if it fails, all other processes are likely to be ineffective" (p. 245).

Two components must be addressed in developing a risk assessment—probability of occurrence and severity of outcome. Hazard identification and analysis establishes severity—the probable harm or damage that could result if an incident occurs. To convert a hazard analysis into a risk assessment, a probability of occurrence factor must be added. Then, risk levels can be established (e.g., low, moderate, serious, high) and priorities can be set.

A hazard is defined as the potential for harm. Hazards include all aspects of technology and activity that produce risk. Hazards are the generic base of, as well as the justification for, the existence of the practice of safety. If no hazards—no potential for harm—existed, safety professionals need not exist. The entirety of purpose of those responsible for safety, regardless of their titles, is to manage with respect to hazards so that the risks deriving from them are acceptable.

Thus, the case can be soundly made that risk assessment should be the core of an operational risk management system. In the author's experience, if workers at all levels have more knowledge and awareness of hazards and risks, fewer serious injuries and fatalities will occur. Getting the required knowledge embedded into workers' minds requires a major, ongoing endeavor. Specifically directed communication and training must be crafted to achieve the awareness and knowledge required—and to achieve the necessary culture change.

Risk assessment literature is abundant. For example, ANSI/ASSE Z690.3, Risk Assessment Techniques, reviews 31 techniques such as primary hazard analysis, fault tree analysis, hazard and operability studies, bow tie analysis, Markov analysis and Bayesian statistics. Uncomplicated systems that could be introduced to supervisors and frontline employees are not as prevalent. However, such a system is contained in an extension of the EU bulletin as a five-step approach:

1) Identify hazards and those at risk.
2) Evaluate and prioritize risks.
3) Decide on preventive action.
4) Take action.
5) Monitor and review results.

Empowering employees to competently assess risks and encouraging them to adopt a mind-set whereby identifying and analyzing hazards and their risks become integral to how they approach and think about work would be a major step forward in injury and fatality prevention. Having knowledge of hazard identification and analysis and risk assessments become rooted within an organization's culture is the type of innovative action needed to further reduce serious injury and fatality potential.

Prevention Through Design

Australia's Guidance on the Principles of Safe Design for Work discusses the contribution of machinery and equipment design to that country's fatality and injury rate. "Of the 210 identified workplace fatalities, 77 (37%) definitely or probably had design-related issues involved. Design contributes to at least 30% of work-related serious nonfatal injuries" (p. 6). The author's review of incident investigation reports (not limited to machinery) concluded that more than 35% had implications of workplace and work methods design inadequacies.

Therefore, to reduce serious injury and fatality potential, prevention through design (PTD) should be established as a separately identified element within an operational risk management system. To help educate designers, safety professionals can develop supportive data on incidents in which design shortcomings were identified and undertake a major effort to have ANSI/ASSE Z590.3-2011, Prevention Through Design: Guidelines for Addressing Occupational Hazards and Risks in Design and Redesign Processes, accepted as a design guide.

> Z590.3 Scope: This standard provides guidance on including prevention through design concepts within an occupational safety and health management system. Through the application of these concepts, decisions pertaining to occupational hazards and risks can be incorporated into the process of design and redesign of work premises, tools, equipment, machinery, substances and work processes including their construction, manufacture, use, maintenance, and ultimate disposal or reuse. This standard provides guidance for a life-cycle assessment and design model that balances environmental and occupational safety and health goals over the life span of a facility, process or product.

Z590.3 says the goal is, as far as is practicable, to ensure that the design selected meets these criteria:

- An acceptable risk level, as defined in this standard is achieved.
- The probability of personnel making human errors because of design inadequacies is at a practical minimum.
- The ability of personnel to defeat the work system and prescribed work methods is at a practical minimum.
- Prescribed work processes consider human factors (ergonomics)—the capabilities and limitations of the work population.
- With respect to access and maintenance, hazards and risks are at a practical minimum.
- The need for PPE is at a practical minimum, and aid is provided for its use where it is necessary (e.g., anchor points for fall protection).
- Applicable laws, codes, regulations and standards have been met.
- Any recognized code of practice, internal or external, has been considered.

Proposing that PTD be a specifically defined element in an operational risk management system is also influenced by ongoing transitions in the methods to eliminate or reduce the occurrence of human error.

Human Error Prevention

During ASSE's "Rethink Safety: A New View of Human Error and Workplace Safety" symposium, speakers commented on topics such as cognitive theory, properties of human cognition, variable errors and constant errors, imperfect rationality and mental behavioral aspects of error. Regarding the sources of human error and corrective actions, some of the commentary was surprising. For example:

• The first step to be taken when human errors occur is to examine the design of the workplace and the work methods.

• Managers may wish to address human error by "getting into the heads" of their employees with training being the default corrective action; training will not be effective if error potential is designed into the work.

• It is management's responsibility to anticipate errors and to have work systems and work methods designed so as to reduce error potential.

Given this, SH&E professionals should study the specifics of a particular situation and the types of errors that may occur, such as those involving complacency when high-hazard tasks are performed repetitively. Dekker (2006) provides insight into what is happening in the human error field. Several excerpts from *The Field Guide to Understanding Human Error* follow.

> Human error is not a cause of failure. Human error is the effect, or symptom, of deeper trouble. Human error is . . . systematically connected to features of people's tools, tasks, and operating systems. (p. 15)

> Sources of error are structural, not personal. If you want to understand human error, you have to dig into the system in which people work. You have to stop looking for people's shortcomings. (p. 17)

> "Rather than being the main instigator of an accident, operators tend to be the inheritors of system defects created by poor design, incorrect installation, faulty maintenance and bad management decisions. Their part is usually that of adding the final garnish to a lethal brew whose ingredients have already been long in the cooking." (p. 88, citing Reason, 1990, p. 173)

> The systemic accident model . . . focuses on the whole [system], not [just] the parts. It does not help you much to just focus on human errors, for example, or an equipment failure, without taking into account the sociotechnical system that helped shape the conditions for people's performance and the design, testing and fielding of that equipment. (p. 90)

> System accidents result not from component failures, but from inadequate control or enforcement of safety-related constraints on the development, design and operation of the system. (p. 91)

This transition in the human error field-—moving from a focus on attempting to change worker behavior to an emphasis on improving the design of the system in which people work—also supports the premise that PTD be a specifically defined element in an operational risk management system.

Management of Change/Prejob Planning

Management of change (MOC)/prejob planning is a process to be applied before modifications are made and continuously throughout the modification activity to ensure that:
- hazards are identified and analyzed, and risks are assessed;
- appropriate avoidance, elimination or control decisions are made so that acceptable risk levels are achieved and maintained during the change process;
- new hazards are not knowingly created by the change;
- the change does not have a negative effect on previously resolved hazards;
- the change does not make the potential for harm of an existing hazard more severe.

In the MOC process, safety professionals should consider the safety of employees making the changes, employees in adjacent areas and those who will be engaged in operations after changes are made. Other considerations include environmental issues, public safety, product safety and quality factors, and avoiding property damage and business interruption.

The author's review of more than 1,700 incident investigation reports, mostly for serious injuries and fatalities, supports the need for and the benefit of MOC systems. These reports showed that a significantly large share of the incidents occurs:
- during unusual and nonroutine work;
- in nonproduction activities;
- in at-plant modification or construction operations (e.g., replacing a motor weighing 800 lb on a platform 15 ft above the floor);
- during shutdowns for repair and maintenance, and startups;
- where sources of high energy are present (electrical, steam, pneumatic, chemical);
- where upsets occur (situations going from normal to abnormal).

Having an effective MOC system would reduce the probability of serious injuries and fatalities occurring in these operational categories.

Petersen (1998) was an early promoter of giving particular attention to serious injury prevention.

> If we study any mass data, we can readily see that the types of accidents that result in temporary total disabilities are different from the types of accidents resulting in permanent partial disabilities or in permanent total disabilities or fatalities. The causes are different. There are different sets of circumstances surrounding severity. Thus, if we want to control serious injuries, we should try to predict where they will happen. Today, we can often do just that. (p. 12)

The key to Petersen's message is prediction. The prior list of work activities was based on reviewing injury and fatality reports. Each entity should develop its own list based on inherent risks and history.

Tom Krause (personal communication) provides additional and substantial support for having an MOC system. Seven companies participated in a 2011 study. Incidents with serious injury or fatality potential were separated from the reports collected. Prejob planning shortcomings were noted in 29% of the incidents that had serious injury or fatality potential. For the nonserious injury potential group, these inadequacies were identified in 17%.

Experience in the auto industry also highlights the benefits of an MOC system. A United Auto Workers (UAW) bulletin covering 1973 through 2007 (no longer available) indicated that 42% of fatalities involved skilled-trades workers, who made up about 20% of the membership. UAW personnel provided data in 2012 (via personal e-mail and phone correspondence) indicating that from 2008 through 2011, 47% of fatalities involved skilled-trades workers. Skilled-trades workers in every industry are often involved in unusual and nonroutine work, at-plant modification or construction operations, shutdowns for repair and maintenance, startups and where sources of high energy are present. The data clearly establish that having an MOC system in place as an element within an operational risk management system can diminish the potential for serious injuries and fatalities.

Incident Investigation

Incident investigation reports may include valuable data on predictive indicators pertaining to serious injury and fatality potential. But, the gap between issued investigation procedures and what actually takes place can be huge. Even in the best safety management systems, incident investigation can be low quality. For example, one large organization determined that if its safety staff promoted adoption of a system as uncomplicated as the five-why technique to improve incident investigation and achieved a B+ grade in 2 years it would be an astounding result.

Therefore, incident investigations should be assessed to identify areas for improvement. Such evaluations often indicate that culture problems exist and that it has become accepted practice for supervisors, managers and safety practitioners to sign-off on shallow investigation reports.

Because of the significance of the information that can be produced, incident investigations must be high quality to reduce the potential for serious injuries and fatalities. Successful investigation systems for near misses that could have resulted in serious results in slightly differing circumstances may also produce critical predictive indicators.

Macro Thinking: The Sociotechnical Concept

Taking a macro view of systems as a whole and adopting the sociotechnical concept (Figure 1) will advance the state of the art in the practice of safety. Several writers say that the term *sociotechnical* was coined in the 1960s by Eric Trist and Fred Emery, then consultants at the Tavistock Institute in London. Researchers at Tavistock said that, based on their research and experience, highly effective operations require a good fit between an organization's technical subsystems and its social subsystems.

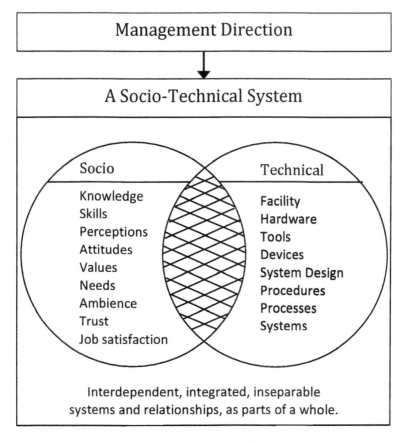

Figure 1: Sociotechnical System

According to Trist and Emery, the interdependency of the technical and social subsystems must be recognized in the design process. Thus, the decision makers would be aware of the effect each subsystem has on the other and design accordingly to ensure that the subsystems work in harmony.

Although the term *sociotechnical systems* is not prominent in the current literature about the organization of work, the idea it conveys is dominant in conventional thinking about the interrelationship between the technical and social aspects of operations.

As noted, Dekker (2006) speaks of "the sociotechnical system":

> The systemic accident model . . . focuses on the whole [system], not [just] the parts. It does not help you much to just focus on human errors, for example, or an equipment failure, without taking into account the sociotechnical system that helped shape the conditions for people's performance and the design, testing and fielding of that equipment. (p. 90)

Dekker explains that operating systems should be considered and examined as a whole to properly determine solutions to problems. Definitions of a sociotechnical system vary in detail although they maintain the substance of what the originators of the term intended. The following definition is a composite that takes a holistic view:

> A sociotechnical system stresses the holistic, interdependent, integrated and inseparable relationship between humans and machines and fosters the shaping of both the technical and the social conditions of work in such a way that both the output goal of the system and the needs of [workers] are accommodated.

The technical aspects of a system include the facility, hardware, tools, devices, system design, physical surroundings and prescribed procedures. The social system consists of the knowledge and skills of employees, from the most senior management level to the newly hired; the attitudes that derive from their beliefs, values, needs, job satisfaction, respect, trust, relations with each other, workplace spirit and ambience, authority structures and reward system; and whether the site has an open communications system through which the views of all can be heard.

When seeking to improve an operational risk management system, SH&E professionals should consider a systems approach to determining the specifics of the recommendation and how the improvement may affect other operational aspects. In taking a sociotechnical systems approach, several concepts should be understood:

- An organization's technical and social systems are inseparable parts of a whole.
- The parts are interrelated and integrated.
- Changing one system may affect others.
- The organization and its employees are not well served if, when resolving a risk situation, the subject at hand is considered in isolation rather than as part of an overall system.

Through this approach, emphasis is given to the importance of the whole as an integrated system and the interdependence of its parts. A feedback process would be created to monitor alignment.

The Sociotechnical System and Safety Culture

Many definitions of safety culture include terms such as shared beliefs, attitudes, values and norms of behavior; shared assumptions; individual and group attitudes about safety; and entrenched attitudes and opinions which a group of people share. This author has used such terms. Note the frequent use of the term *shared*.

However, such definitions need examination. An organization's safety culture, which is a subset of its overall culture, derives from the decisions made at the executive level. An organization's culture with respect to occupational safety, environmental safety, public safety and product safety is determined by the outcome of management decisions as measured by the risk levels in a facility's technical and social aspects. Management owns the culture. Employees may or may not share the views and beliefs held by management with respect to safety and operational risk levels. For example, employees may believe that an operation is overly risky while management accepts the risk level. The culture created by management is the dominant factor with respect to the risk levels attained, acceptable or unacceptable.

Conclusion

The safety profession must consider adopting a systemic sociotechnical model for an operational risk management system as a substantially different means to improve serious injury and fatality prevention.

The board of directors and senior management establish a culture that requires maintaining acceptable risk levels in all operations.

Management leadership, commitment, involvement and the accountability system, establish that the performance level to be achieved is in accord with the culture established by the board.

To achieve acceptable risk levels, management establishes policies, standards, procedures and processes with respect to:

Providing adequate resources
Risk assessment, prioritization and management
  •Applying a hierarchy of controls
Prevention through design
  •Inherently safer design
  •Resiliency
Maintenance for system integrity
Competency and adequacy of staff
  •Capability—skill levels
  •Sufficiency in numbers
Safety-related systems
  •Training—motivation
  •Employee participation
  •Information—communication
  •Work methods, scheduling
  •Permit systems

- Inspections
- Incident investigation and analysis
- PPE

Management of change/prejob planning
Third-party services
- Relationships with suppliers
- Safety of contractors—on premises

Procurement—safety specifications
Emergency planning and management
Compliance assurance

Performance measurement: Evaluations are made and reports are prepared for management review to support continuous improvement and to ensure that acceptable risk levels are maintained.

In this model, sociotechnical concepts should be foundational:

- at a board level, where establishing a culture begins;
- in the decisions made at a senior management level to demonstrate its involvement to achieve the culture and acceptable risk levels;
- in policies, standards, procedures and processes established;
- throughout the administration of all of the individual aspects of the operational risk management system.

It is close to the top level on a probability scale that acceptable risk levels can be achieved and maintained if an operational risk management system is built on this model and that superior financial results would be achieved as well. Effectively adopting the elements in this model will reduce serious injury and fatality potential. It will have the same effect on the occurrence of all other injuries.

Safety professionals are encouraged to take elements from this model, sequentially, that they believe can be fit into the management systems in place and which are compatible with an organization's culture. That said, each element in this model should be included in the ideal for an operational risk management system.

The model presented is a major departure from other outlines for a safety management system. However, it supports the position that "major and somewhat drastic innovations in the content and focus of occupational risk management systems will be necessary to achieve additional progress in serious injury and fatality prevention." ■

References

Aldridge, J. (2004). Sociotechnical issues. In A. Distefano, K. Rudestam & R. Silverman (Eds.), *Encyclopedia of distributed learning* (pp. 413-417). Thousand Oaks, CA: SAGE Publications. Retrieved from www.argo-s-press.com/Resources/team-building/socitechnisistem.htm

ANSI/Association for Manufacturing Technology (AMT). (2010). Safety of machinery: General safety requirements and risk assessments (ANSI/AMT B11.0). Leesburg, VA: Author.

ANSI/AIHA/ASSE. (2012). Occupational health and safety management systems (ANSI/ASSE Z10-2012). Des Plaines, IL: Author.

ANSI/ASSE. (2011). Prevention through design: Guidelines for addressing occupational hazards and risks in design and redesign processes (ANSI/ASSE Z590.3-2011). Des Plaines, IL: Author.

ANSI/ASSE. (2011). Risk assessment techniques (ANSI/ASSE Z690.3). Des Plaines, IL: Author.

Australian Safety and Compensation Council. (2006). Guidance on the principles of safe design for work. Canberra, Australia: Author.

British Standards Institute (BSI). (2007). Occupational health and safety management systems: Requirements (BS OHSAS 18001:2007). London, UK: Author.

Bureau of Labor Statistics (BLS). Manufacturing employment. Retrieved from http://data.bls.gov/timeseries/CES3000000001?data_tool=XGtable

BLS. (2011). Census of fatal occupational injuries, 1996-2011. Retrieved from www.bls.gov/iif

BLS. (2001). The employment situation: January 2011. Retrieved from www.bls.gov/news.release/archives/empsit_02042011.pdf

Chapanis, A. (1980). The error-provocative situation. In W. Tarrants, *The measurement of safety performance*. New York, NY: Garland Press Publishing.

Davis, J. & Bar-Chaim, Y. (2011). Workers' compensation claims frequency. Retrieved from www.ncci.com/documents/2011_Claim_Freq_Research.pdf

Dekker, S. (2006). *The field guide to understanding human error*. Burlington, VT: Ashgate Publishing Co.

Dwyer, C. (2011). Sociotechnical systems theory and environmental sustainability. *Sprouts: Working Papers on Information Systems, 11*(3). Retrieved from http://sprouts.aisnet.org/11-3

European Agency of Safety and Health at Work. (2008). Risk assessment. Retrieved from http://osha.europa.eu/en/topics/riskassessment

Frey, W. (2013). Sociotechnical systems in professional decision making. Inside collection course: Corporate governance. Retrieved from http://cnx.org/content/m14025/latest/?collection=col10396/latest

Federal Railroad Administration (FRA). (2009). Risk reduction program (49 CFR Chapter II; FRA-2009-0038; RIN 2130-AC11. Washington, DC: DOT, Author.

Ferguson, L. (2007). Forum on fatality prevention [Press release]. Indiana, PA: Indiana University of Pennsylvania.

Health and Safety Executive. (2008). Five steps to risk assessment. London, U.K.: Author.

Heinrich, H.W. (1950). *Industrial accident prevention* (3rd ed.). New York, NY: McGraw-Hill Book Co.

iSixSigma. Determine the root cause: 5 whys. Retrieved from www.isixsigma.com/tools-templates/cause-effect/determine-root-cause-5-whys

Johnson, W. (1980). *MORT safety assurance systems.* New York, NY: Marcel Dekker.

Manuele, F.A. (2008). *Advanced safety management: Focusing on Z10 and serious injury prevention.* Hoboken, NJ: John Wiley & Sons Inc.

Manuele, F.A. (2011, 10). Reviewing Heinrich: Dislodging two myths from the practice of safety. *Professional Safety, 56*(10), 52-61.

Mealy, D. (2005, 2006, 2009). State of the workers' compensation line. Boca Raton, FL: National Council on Compensation Insurance (NCCI).

NCCI. (2005). Workers' compensation claim frequency down again [News bulletin]. Boca Raton, FL: Author.

National Safety Council (NSC). (1995). *Accident facts.* Itasca, IL: Author.

Petersen, D. (1998). *Safety management* (2nd ed.). Des Plaines, IL: ASSE.

Reason, J. (1990). *Human error.* New York, NY: Cambridge University Press.

Reason, J. (1997). *Managing the risks of organizational accidents.* Burlington, VT: Ashgate Publishing Co.

Socio-technical theory. Retrieved from http://istheory.byu.edu/wiki/Socio-technical theory

Trist, E.L. (1981). *The evolution of sociotechnical systems: A conceptual framework and an action research program.* Toronto, Canada: Ontario Quality of Working Life Center.

United Auto Workers (UAW). (2007). Facts and commentary about fatalities in UAW-represented workplaces in 2007." Detroit, MI: Author.

U.S. Census Bureau. All sectors: Geographic area series: Economy-wide key statistics: 2007. Retrieved from http://factfinder2.census.gov/faces/tableservices/jsf/pages/product view.xhtml?src=bkmk

13 ANSI/AIHA/ASSE Z10-2012
AN OVERVIEW OF THE OCCUPATIONAL HEALTH AND SAFETY MANAGEMENT SYSTEMS STANDARD

ORIGINALLY PUBLISHED APRIL 2014

On July 25, 2005, ANSI approved a new standard titled Occupational Health and Safety Management Systems (OHSMS). Its designation was ANSI/AIHA Z10-2005. That was a major development. For the first time in the U.S., a national consensus standard was issued for a safety and health management system applicable to organizations of all sizes and types.

Per ANSI requirements, standards are reviewed every 5 years to be revised or reaffirmed. AIHA, then-secretariat, formed a committee to review Z10. The outcome of its work was a revised standard approved on June 27, 2012. Its designation was ANSI/AIHA Z10-2012. Shortly after approval, secretariat was transferred to ASSE.

All persons who give counsel on occupational safety and health management systems should own a copy of this standard and be thoroughly familiar with its content. The 2012 version reflects significant changes and contains valuable support information in the advisory column and the appendixes.

The standard provides senior management with a well-conceived concept and action outline for a safety and health management system. As employers make changes to meet the standard's requirements, it can be expected that occupational injuries and illnesses will be reduced.

To identify differences and to develop a prioritized improvement plan, SH&E professionals should conduct a gap analysis to compare the elements in existing safety programs, processes or systems with the requirements in Z10. That comparison should be followed by a prioritized action plan for continual improvement.

Participation: Consensus

More than 40 safety professionals served on the committee that crafted the 2012 version of Z10. They represented industry, labor, government, business associations, professional organizations, academe and persons of interest.

The Z10 committee adhered strictly to the due diligence requirements applicable to development of an ANSI standard. A balance of stakeholders provided input and open discussion, which resulted in the group vetting each issue that was raised to an agreed-upon conclusion.

In crafting the current version of Z10, the intent was to present management system requirements that when effectively implemented would not only achieve significant safety and health benefits, but also have a favorable effect on productivity, financial performance, quality and other business goals. The standard is built on the well-known plan-do-check-act (PDCA) process for continual improvement (Figure 1).

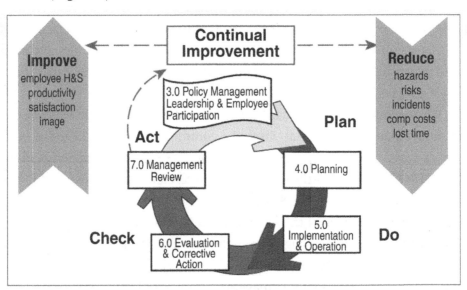

Figure 1: Plan-Do-Check-Act Concept
The Z10-2012 standard is built on the well-known plan-do-check-act process for continual improvement. It provides senior management with a well-conceived concept and action outline for a safety and health management system.

Many companies have issued safety policy statements in which they indicate their intent to comply with or exceed all relative laws and standards. Those employers, particularly, will want to implement the Z10 provisions that are not part of their current safety and health management systems.

A Major Theme

Throughout all sections of Z10, from management leadership and employee participation through the management review provisions, a key theme is prominent: Processes for continual improvement are to be in place and implemented to ensure that:

- hazards are identified and evaluated;
- risks are assessed and prioritized;
- management system deficiencies and opportunities for improvement are identified and addressed;
- risk elimination, reduction or control measures are taken to ensure that acceptable risk levels are attained.

In relation to this theme, the following terms as defined by the standard are particularly applicable.

> **Hazard:** A condition, set of circumstances or inherent property that can cause injury, illness or death.
>
> **Exposure:** Contact with or proximity to a hazard, taking into account duration and intensity.
>
> **Risk:** An estimate of the combination of likelihood of an occurrence of a hazardous event or exposure(s) and severity of injury or illness that may be caused by the event or exposures.
>
> **Probability:** The likelihood of a hazard causing an incident or exposure that could result in harm or damage for a selected unit of time, events, population, items or activity being considered.
>
> **Severity:** The extent of harm or damage that could result from a hazard-related incident or exposure.
>
> **Risk assessment:** Process(es) used to evaluate the level of risk associated with hazards and system issues. (ANSI/AIHA/ASSE, 2012)

The definitions are repeated in Appendix F, which provides guidance on risk assessment. Although *acceptable risk* is not a term included in the standard's definitions, it is made clear in several places that the goal is to achieve acceptable risk levels. For example, Section 6.4, Corrective and Prevention Actions, clearly states that an organization is to have processes in place to ensure that acceptable risk levels are achieved and maintained. Appendix F states, "The goal of the risk assessment process including the steps taken to reduce risk is to achieve safe working conditions with an acceptable level of risk."

Z10 Is a Management System Standard

Z10 is a management system standard, not a specification standard. What is the difference? A management system standard provides process and system guidelines for a provision without specifying the details on how the provision is to be carried out. A specification standard contains those details.

Z10 Section 5.2B illustrates the difference:

> Section 5.2: Education, Training, Awareness and Competence. The organization shall establish processes to:
> B. Ensure through appropriate education, training or other methods that employees and contractors are aware of applicable OHSMS requirements and their importance are competent to carry out their responsibilities as defined in the OHSMS. (ANSI/AIHA/ASSE, 2012)

That is the extent of the "shall" requirements for Section 5.2B. The explanatory part of the standard contains comments on subjects for which training should be given, such as safety design, incident investigation, hazard identification, good safety practices and the use of PPE. However, those comments are advisory and not a part of the standard.

If Z10 were written as a specification standard, requirements comparable to the following might be extensions of 5.2B.

a) At least 12 hours of training shall be given initially to engineers and safety professionals in safety through design, to be followed annually with a minimum of 6 hours of refresher materials.

b) All employees shall be given a minimum of 3 hours training annually in hazard identification.

c) All employees shall be given a minimum of
4 hours training annually in the use of PPE.

d) All training activities conducted as a part of this provision shall be documented and records shall be retained for a minimum of 5 years.

Compatibility and Harmonization

One goal of the Z10 committee was to ensure that it could be easily integrated into any management system an organization has in place. That goal was met. As to structure, the standard is compatible and harmonized with quality and environmental management system standards—the ISO 9000 and ISO 14000 series at the time of the approval of Z10.

Employment Implications

A brief verbal survey of safety degree program professors was conducted to determine what qualifications were being stressed by human relations personnel when they visited campuses to recruit. The survey found that they want candidates who are equipped with the knowledge and skill to give counsel on many of the provisions in Z10. In that respect, safety professionals should give particular attention to certain provisions. Those provisions are in sections: 4.0, Planning; 5.0, Implementation and Operations; and 6.0, Evaluation and Corrective Action.

In summary, those sections state that employers shall establish and implement processes to:

•Identify and control hazards in the design process and when changes are made in operations. This requires safety design reviews for new and altered facilities and equipment.

•Have an effective management of change system in place through which hazards and risks are identified and evaluated in the change process.

•Assess the level of risk for identified hazards for which knowledge of risk assessment methods will be necessary.

•Utilize a prescribed hierarchy of controls in dealing with hazards to achieve acceptable risk levels for which the first step is to attempt to design out or otherwise eliminate the hazard.

•Avoid bringing hazards into the workplace by incorporating design and material specifications into procurement contracts for facilities, equipment and materials.

Certification Implications

Provisions in Z10 have a direct relationship to the content of the CSP examinations. Those examinations are reviewed about every 5 years to ensure that they are current with respect to what safety professionals actually do. In that review process, safety professionals are asked to describe the content of their work at the time the survey is made.

The author compared the content of the Comprehensive Practice Examination Guide issued by BCSP in 2011 with that issued in 2006. Substantial changes were made in the later edition, many of which relate to principal requirements of Z10. For example, a domain in the sixth edition is devoted entirely to hazard identification and analysis and risk assessment. It is the centerpiece of the examination. Knowledge requirements for safety through design and management of change are more extensive as well.

Z10 represents sound, current practice. Having knowledge of and experience with its provisions is required to pass the CSP examination.

Continual Improvement: The PDCA Concept

Z10 is built on the well-known PDCA process for continual improvement. Understanding this process is necessary to effectively implement the standard. The introduction to Z10 includes a chart based on the PDCA concept that emphasizes continual improvement.

Throughout the standard, the words *process, processes, systems* and *continual improvement* are often repeated. Z10 emphasizes a continual improvement approach. As noted, the standard outlines the processes to be put in place to have an effective safety and health management system, not the specifics. Each organization must determine process specifics based on its unique hazards and risks.

A Review of the Standard's Provisions

Brief comments are made here to provide an overview of the standard's major sections. With respect to these remarks, keep in mind that Z10 is presented in two columns. The standard is in the left column; the explanatory and advisory material is in the right column—the E column.

As is common in ANSI standards, requirements are identified by the word *shall*. An organization that chooses to conform to the standard is expected to fulfill the shall requirements. The word *should* is used to describe recommended practices or give an explanation of the requirements.

Numerous recommended practices and advisory comments are included in the E column and in the appendices to assist in the implementation of the standard; they are not requirements.

Z10-2012 Table of Contents

The Z10-2012 table of contents provides a base for review and comparison with the safety management systems with which safety practitioners are familiar.

Chapter 13: ANSI/AIHA/ASSE Z10-2012: An Overview of the OHSMS Standard

Table of Contents
Foreword
1.0 Scope, Purpose and Application
   1.1 Scope
   1.2 Purpose
   1.3 Application
2.0 Definitions
3.0 Management Leadership and Employee Participation
   3.1 Management Leadership
      3.1.1 Occupational Health and Safety Management System
      3.1.2 Policy
      3.1.3 Responsibility and Authority
   3.2 Employee Participation
4.0 Planning
   4.1 Initial and Ongoing Reviews
   4.2 Assessment and Prioritization
   4.3 Objectives
   4.4 Implementation Plans and Allocation of Resources
5.0 Implementation and Operation
   5.1 OHSMS Operational Elements
      5.1.1 Risk Assessment
      5.1.2 Hierarchy of Controls
      5.1.3 Design Review and Management of Change
      5.1.4 Procurement
      5.1.5 Contractors
      5.1.6 Emergency Preparedness
   5.2 Education, Training, Awareness and Competence
   5.3 Communication
   5.4 Document and Record Control Process
6.0 Evaluation and Corrective Action
   6.1 Monitoring, Measurement and Assessment
   6.2 Incident Investigation
   6.3 Audits
   6.4 Corrective and Preventive Actions
   6.5 Feedback to the Planning Process
7.0 Management Review
   7.1 Management Review Process
   7.2 Management Review Outcomes and Follow Up
Appendixes
   A) Policy Statements (Section 3.1.2)
   B) Roles and Responsibilities (Section 3.1.3)
   C) Encouraging Employee Participation (Section 3.2)
   D) Planning-Identification, Assessment and Prioritization (Section 4.0)
   E) Objectives/Implementation Plans (Section 4.3 and 4.4)
   F) Risk Assessment (Section 4.1 and 5.1.1)
   G) Hierarchy of Control (Section 5.1.2)
   H) Management of Change (Section 5.1.3)
   I) Procurement (Section 5.1.4)

J) Contractor Safety and Health (Section 5.1.5)
K) Incident Investigation Guidelines (Section 6.2)
L) Audit (Section 6.3)
M) Management Review Process (Section 7.1 and 7.2)
N) Management System Standard Comparison (Introduction)
O) Bibliography and References

New appendixes in the 2012 version are: F) Risk Assessment; I) Procurement; J) Contractor Safety and Health; M) Management of Change; N) Management System Standard Comparison. While the appendixes are not part of the standard, they can be helpful to those with implementation responsibility.

Section 1.0 Scope, Purpose and Application

Section 1.1 Scope defines the minimum requirements for OHSMS. According to the advisory data for the scope, the intent is to provide a systems approach for continual improvement in safety and health management, and to avoid specifications. Furthermore, the writers recognized the uniqueness of the culture and organizational structures of individual organizations and the need for each entity to "define its own specific measures of performance."

Section 1.2 Purpose indicates the primary purpose of the standard: To provide a management tool to reduce the risk of occupational injuries, illnesses and fatalities.

Section 1.3 Application states that the standard is applicable to organizations of all sizes and types. Z10 contains no limitations or exclusions by industry, business type or number of employees. The introduction and advisory column comments opposite Section 1.3 state that the standard's structure allows integration with quality and environmental management systems. Doing so is a good idea.

Section 2.0 Definitions

As is typical in ANSI standards, certain terms are defined as used in the standard. Safety professionals should become familiar with them.

Section 3.0 Management Leadership and Employee Participation

Section 3.0 is the standard's most important section. Safety professionals will surely agree with the premise that top management leadership and effective employee participation are crucial for OHSMS success. Top management leadership is vital. An organization's safety culture derives from decisions made at that level. And, continual improvement cannot be achieved without effective direction from management. Key statements in the shall column of the standard follow:

3.1.1 Top management shall direct the organization to establish, implement and maintain an OHSMS.

3.1.2 The organization's top management shall establish a documented occupational health and safety policy.

3.1.3 Top management shall provide leadership and assume overall responsibility.

3.2 The organization shall establish a process to ensure effective participation in the OHSMS by its employees at all levels.

As management provides direction and leadership, assumes responsibility for the OHSMS and ensures effective employee participation, it is important to keep the standard's purpose in mind: To reduce the risk of occupational injuries, illnesses and fatalities. That will be best accomplished if personnel understand that achieving acceptable risk levels is the desired outcome of every safety and health management process. Appendixes A, B and C provide supporting data on policy statements, roles and responsibilities, and employee participation.

Section 4.0 Planning

In the PDCA process, planning is the first step. As would be expected, Section 4.0 Planning sets forth the planning process to implement the standard and to establish improvement plans. The goal is to identify and prioritize issues within a safety management system that need improvement. Those issues are defined as hazards, risks and management system deficiencies. Throughout, the standard emphasizes having systems and processes in place to identify hazards and assess their accompanying risk and to identify the management deficiencies related to them.

In the continual improvement process, information that defines opportunities for further improvement in the OHSMS and, thereby, risk reduction, is fed back into the planning process for additional consideration.

Section 4.1 requires that a company conduct a comprehensive review to identify the differences between existing systems and the requirements of the standard. The review shall encompass: business and operational processes that are relative to the standard's requirements; operational issues mentioned in the planning section (hazards, risks and management system deficiencies); and allocation of the resources necessary to achieve and maintain an acceptable risk level; applicable regulations and standards; content of risk assessments made; management system audits; and, as emphasized in the standard, the means established for employee participation and contribution.

Section 4.2 sets forth the requirements for assessment and prioritization.

The organization shall establish a process to assess and prioritize OHSMS issues on an ongoing basis. The process shall:

A) Assess the impact on health and safety of OHSMS issues, and assess the level of risk for identified hazards;

B) Establish priorities based on factors such as the level of risk, potential for system improvement, standards, regulations, feasibility and potential business consequences;

C) Identify underlying causes and other contributing factors related to system deficiencies that lead to hazards and risks.

Following are excerpts from selected explanatory notes in sections 4.2A and 4.2B.

At E4.2A, the standard's writers offer advice on risk assessments and the factors that they should include, such as potential hazards and exposures; frequency of exposure; human behavior; controls in place and their effectiveness; and the *potential severity of hazards* (emphasis added for the last point).

At E4.2B, it is made clear the business results that may relate to the standard's application may include either increased or decreased productivity, sales, net income or public image.

Thus, employers need processes to identify and analyze hazards, assess the risks deriving from those hazards, and establish mitigation priorities which, when acted on, will attain acceptable risk levels. Appendix D provides guidance on assessment and prioritization.

4.3 Objectives, says: "The organization shall establish a process to set documented objectives, quantified where practicable, based on issues that offer the greatest opportunity for OSHMS improvement and risk reduction."

4.4 Implementation Plans and Allocation of Resources follows logically in accord with a sound problem-solving procedure. After hazards, risks and shortcomings in safety management systems have been identified and objectives have been outlined, a documented plan must be established and implemented to achieve the objectives.

Item B in Section 4.4 says that an employer must assign adequate resources to achieve the objectives outlined in the implementation plans. It is an absolute that if adequate resources are not provided, acceptable risk levels cannot be maintained.

Section 5.0 Implementation and Operation

This section defines the operational elements that are required for implementation of an effective OHSMS. These elements provide the backbone of an OHSMS and the means to pursue the objectives from the planning process. Note the phrase "the backbone of an OHSMS."

5.1 OHSMS Operational Elements are to be integrated into the management system. A new and important addition to Z10 appears in this section.

Section 5.1.1 Risk Assessment

"The organization shall establish and implement a risk assessment process(es) appropriate to the nature of hazards and level of risk."

Adding this shall provision reflects a worldwide trend emphasizing the importance of risk assessments. Appendix F provides a six-page overview of risk assessment and includes data on several techniques. Having knowledge of preliminary hazards analysis, what-if/checklist analysis, and failure modes and effects analysis and how they are applied will satisfy most needs of safety professionals as they give counsel on risk assessment.

In applying hazard analysis and risk assessment techniques, SH&E professionals may use qualitative and quantitative methods. Mathematical calculations required will not be extensive.

Appendix F also gives an example of a risk assessment matrix. Most risk assessment matrixes set forth incident probability categories, severity of harm or damage ranges, and resulting risk levels. Such a matrix can serve as a valuable instrument when working with decision makers to set risk levels and prioritize corrective actions. Variations in published risk assessment matrixes are substantial. A safety professional should customize the matrix to the organization's needs.

Section 5.1.2 Hierarchy of Controls

Provisions for the use of a specifically defined hierarchy of controls are outlined. The organization shall apply the methods of risk reduction in the order prescribed. This is how the standard and the explanatory comments read.

> The organization shall establish a process for achieving feasible risk reduction based upon the following preferred order of controls:
> A) elimination;
> B) substitution of less hazardous materials, processes, operations or equipment;
> C) engineering controls;
> D) warnings;
> E) administrative controls;
> F) personal protective equipment.

Sound management principles shall be applied as decisions are made with respect to the hierarchy of controls. Decision makers should consider the nature of the hazards and their accompanying risks; scope of risk reduction that must be achieved; necessity to adhere to applicable standards and regulations, both external and internal; what is considered good practice in the industry; available technology; and cost effectiveness.

A hierarchy is a system of persons or things ranked one above the other. The hierarchy of controls in Z10 provides a systematic way of considering steps in a ranked and sequential order to select the most effective means of eliminating or reducing hazards and their risks. Acknowledging the premise that risk reduction measures should be considered and taken in a prescribed order represents an important step in the evolution of the practice of safety.

Appendix G provides a pictorial and verbal display of the hierarchy of controls listed in 5.1.1 with application examples for each element.

Section 5.1.3 Design Review and Management of Change

The following excerpts indicate what Z10 requires for design reviews and management of change. To repeat for emphasis, these are shall provisions.

> The organization shall establish a process to identify, and take appropriate steps to prevent or otherwise control hazards at the design and redesign stages, and for situations requiring management of change to reduce potential risks to an acceptable level. The process for design and redesign and management of change shall include:
> A) identification of tasks and related health and safety hazards;
> B) recognition of hazards associated with human factors including human errors caused by design deficiencies;
> C) review of applicable regulations, codes, standards, and internal and external recognized guidelines;
> D) application of control measures (hierarchy of controls, Section 5.1.2);
> E) a determination of the appropriate scope and degree of the design review and management of change;
> F) employee participation.

The Design Process

The author and others have long professed that the most effective and economical way to achieve acceptable risk levels is to have the hazards from which the risks derive addressed in the design process. That is what this standard requires. It is an exceptionally important element in this standard. Impact of its application can be immense.

Management of Change

Employers must have processes in place to identify and prevent or otherwise control hazards and reduce the potential risks associated with them when existing operations, products, services or suppliers change. Getting effective management of change procedures in place is not easy.

The author's research shows that for all occupations many incidents that result in serious injury occur when out-of-the-ordinary situations arise, particularly when unusual and nonroutine work is being performed and when sources of high energy are present (Manuele, 2008, p. 51; 2012, p. 166). Safety professionals should study thoroughly the standard's management of change requirements to determine how they might promote the culture change necessary for their implementation. Applying change management methods will be necessary.

Note that sections 5.1.3.1 and 5.1.3.2 are extensions of 5.1.3 Design Review and Management of Change.

Section 5.1.3.1 Applicable Life Cycle Phases

The provision in Section 5.1.3.1 says that an organization must consider the entirety of the life cycle of the subject being designed or redesigned.

At E5.1.3.1, the standard's writers advise that the design process may apply in all or some of the following: concept stages; preliminary design; detailed design; build or purchase process; commissioning, installing and debugging processes; production and maintenance operations; and decommissioning activity.

Section 5.1.3.2 Process Verification

> The organization shall have processes in place to verify that changes in facilities, documentation, personnel and operations are evaluated and managed to ensure safety and health risks arising from these changes are controlled.

Section 5.1.4 Procurement

Although the requirements for procurement are plainly stated and easily understood, they are brief in relation to the enormity of what will be required to implement them. An interpretation of the requirements could be: "Safety practitioners, you are assigned the responsibility to convince management and purchasing agents that, in the long term, it can be very expensive to buy cheap." This is how the standard and the explanatory data read.

> The organization shall establish and implement processes to:
> A) Identify and evaluate the potential health and safety risks associated with purchased products, raw materials, and other goods and related services before introduction into the work environment;
> B) Establish requirements for supplies, equipment, raw materials, and other goods and related services purchased by the organization to control potential health and safety risks;
> C) Ensure that purchased products, raw materials, and other goods and related services conform to the organization's health and safety requirements.

Appendix I provides guidance on the procurement process. Adding an element to an OHSMS designed to prevent bringing hazards into the workplace could have startling positive results in reducing the frequency and severity of hazardous incidents and exposures.

Section 5.1.5 Contractors

Section 5.1.5 Contractors requires that an organization have processes in place to avoid injury and illness to its employees from activities of contractors and to the contractor's employees from the organization's operations. Many entities have such procedures in place.

One of the shall provisions indicates that the process is to include "contractor health and safety performance criteria." That implies, among other things, vetting a contractor with respect to its previous safety performance before awarding a contract.

Section 5.1.6 Emergency Preparedness

To meet the requirements of the provision in section 5.1.6, an organization must have processes in place to "to identify, prevent, prepare for and/or respond to emergencies." Also, an employer should conduct periodic drills to test the emergency plans, which are then updated.

Section 5.2 Education, Training, Awareness and Competence

An organization is required to determine the knowledge needed to achieve competence; ensure that employees are aware of the OHSMS requirements; remove any barriers to participation in education; ensure that training is ongoing and is delivered in a language trainees understand; and ensure that trainers are competent. This section has six alpha-designated provisions, three of which contain the words *competence* or *competent*. Thus, competence is emphasized. Employees and contractors are to be competent to fulfill their responsibilities. Trainers are to be competent to train.

These provisions are also applicable to contractors. Interestingly, both safety design and procurement are mentioned in the examples of training that should be given. This is how item E5.2A reads.

> E5.2A: Training in OHSMS responsibilities should include, for example, training for: engineers in safety design (e.g., hazard recognition, risk assessment, mitigation, etc.); those conducting incident investigations and audits for identifying underlying OHSMS nonconformances; procurement personnel on impact of purchasing decisions; and others involved with the identification of OHSMS issues, methods of prioritization and controls.

Section 5.3 Communication

An organization is to institute processes to: communicate information about the progress being made on its implementation plan; ensure prompt reporting of incidents, hazards and risks; promote employee involvement so that they make recommendations on hazards and risks; inform contractors and *relevant external interested parties* of changes made that affect them; and remove barriers to all of the foregoing. With respect to contractors, item E5.3D provides guidance as follows:

> E5.3D: The work activities of contractors can pose additional hazards for both employees and others in the workplace. Processes established for consultation with contractors should ensure risks will be appropriately addressed using good OHS practices. This consultation should include discussion and resolution of issues of mutual concern.

Section 5.4 Document and Record Control Process

The standard specifies documentation requirements for certain systems in several places. These processes should fit the requirements of the OHSMS in place. The advice given in the informational column at E5.4 is that the documentation procedures put in place should be commensurate with the size of an organization as represented by the number of employees, the complexity of operations and its inherent hazards and risks.

Documents shall be updated as needed, legible, adequately protected against damage or loss, and retained as necessary.

Section 6.0 Evaluation and Corrective Action

This section outlines the requirements for processes to evaluate the performance of the safety management system, to take corrective action when shortcomings are found and to provide feedback to the planning and management review processes. Communications on lessons learned are to be fed back into the planning process. The expectation is that new objectives and action plans will be written in relation to what has been experienced.

Section 6.1 Monitoring, Measurement and Assessment

> The organization shall establish and implement a process to monitor and evaluate hazards, risks and their controls to assess OHSMS performance. These processes shall include some or all of the following methods, depending on the nature and extent of identified hazards and risks: workplace inspections and testing; assessments of exposures; incident tracking; measuring performance in relation to legal or other requirements; or other methods selected by the organization. Results of the monitoring processes shall be communicated as appropriate.

Section 6.2 Incident Investigation

The standard says:

> The organization shall establish a process to report, investigate and analyze incidents in order to address OHSMS nonconformances and other factors that may be causing or contributing to the occurrence of incidents. The investigations shall be performed in a timely manner.

One advisory comment highlights the value in feeding lessons learned from investigations into the planning and corrective action processes.

Section 6.3 Audits

An organization shall:

> Have audits made periodically with respect to application of the provisions in the OHSMS; ensure that audits are made by competent persons not attached to the location being audited; document and communicate the results; have auditors communicate immediately on potentials for serious injuries or illnesses and fatalities so that swift corrective action can be taken.

Audits are to measure the organization's effectiveness in implementing the OHSMS elements. Thus, audits are to determine whether the management systems in place effectively identify hazards and control risks. Although many safety professionals are familiar with safety audit processes, they should review what the standard requires and determine whether it will be beneficial to revise their audit systems. Appendix L is helpful in this respect; it contains a sample audit protocol that matches all of the sections of Z10.

In the advisory column, the standard's writers make clear that audits are to be system oriented rather than compliance oriented.

Also, and importantly, E6.3B comments on the independence of auditors. While it says that "audits should be conducted by individuals independent of the activities being audited," it also says that "this does not mean that audits must be conducted by individuals external to the organization."

Section 6.4 Corrective and Preventive Actions

Revisions made in the 2012 version of Z10 are substantial in this section. It defines what the organization shall do to fulfill the provisions of this section.

> The organization shall establish and implement corrective and preventive action processes to:
> A) Address nonconformances and hazards that are not being controlled to an acceptable level of risk.
> B) Identify and address new and residual hazards associated with corrective and preventive actions that are not being controlled to an acceptable level of risk.
> C) Expedite action on high-risk hazards (those that could result in fatality or serious injury/illness) that are not being controlled to an acceptable level of risk;
> D) Review and ensure effectiveness of corrective and preventive actions taken.

It is made clear that organizations must identify hazards, risks and shortcomings in management systems, and take corrective action to achieve acceptable risk levels.

Furthermore, the standard says that organizations are to expedite appropriate actions on high-risk hazards.

Section 6.5 Feedback to the Planning Process

Section 6.5 is a communication provision pertaining to all shortcomings in the safety management system. Its purpose is to provide a base for revision in the planning process. The standard says that the communication process established shall ensure a proper flow of the information developed in the monitoring and measurement systems, audits, incident investigations and in the corrective actions taken to those involved in the planning process to achieve continual improvement endeavors.

Section 7.0 Management Review

This section requires that OHSMS performance be reviewed periodically and that management take appropriate actions in response. The management review section and extensive advisory comments pertaining to it are must reads. As noted, the Management Leadership and Employee Participation section is the most important in Z10. The section on management review is a close second. Periodically reviewing management system effectiveness is an important part of the PDCA process.

Section 7.1 Management Review Process

> The organization shall establish and implement a process for top management to review the OHSMS at least annually, and to recommend improvements to ensure its continued suitability, adequacy and effectiveness.
> E7.1: Management reviews are a critical part of the continual improvement of the OHSMS.

Some subjects must be reviewed at least annually: progress in reducing risk; effectiveness of processes to identify, assess and prioritize risk and system deficiencies; effectiveness in addressing underlying causes of risks and system deficiencies; the extent to which objectives have been met; and performance of the OHSMS in relation to expectations.

Section 7.2 Management Review Outcomes and Follow-Up

Section 7.2 requires that management determine whether changes need to be made in "the organization's policy, priorities, objectives, resources or other OHSMS elements to establish the future direction of the OHSMS."

In accord with good management procedures, senior management is expected to give direction to implement the changes needed in OHSMS and processes to continually reduce risks. The standard requires that "results and action items from the management reviews shall be documented and communicated to affected individuals, and tracked to completion." This provision gives the needed importance to the management review process. Action items are to be recorded, communicated to those affected, and followed through to a proper conclusion.

Advisory Content and Appendices

Z10 provides exceptionally valuable explanatory and supportive data in the advisory column and in the appendixes. Alpha-numerical pages 1 through 29 pertain to the standard's requirements and the advisory material. Pages 30 through 88 are devoted to the appendixes. That is about a 65% increase in the space devoted to appendixes compared to the 2005 version. A safety professional must have a copy of the standard to appreciate the value of the guidance material and the appendixes.

Conclusion

This revision to Z10 is important work. Prudent safety professionals will study the standard's requirements to determine whether additional skills and capabilities are needed and take steps to acquire those skills. Having done so, they will be equipped to guide on implementing safety management system elements that may not exist in the organizations to which they give counsel. ■

References

ANSI/AIHA/ASSE. (2012). *American national standard for occupational health and safety management systems* (ANSI/AIHA/ASSE Z10-2012). Fairfax, VA: AIHA; Des Plaines, IL: ASSE.

ANSI/ISO/American Society for Quality (ASQ). (2000). *American national standard for quality management systems—requirements* (ANSI/ISO/ASQ Q9001-2000). Milwaukee, WI: American Society For Quality.

BCSP. (2011, April). *The comprehensive practice examination guide* (6th ed.). Champaign, IL: BCSP.

Christensen, W.C. & Manuele, F.A. (Eds.). (1999). *Safety through design*. Itasca, IL: NSC.

International Organization for Standardization (ISO). (2004). *Environmental management systems—requirements with guidance for use* (ISO 14001:2004). Geneva, Switzerland: Author.

Manuele, F.A. (2008). *Advanced safety management: Focusing on Z10 and serious injury prevention*. Hoboken, NJ: John Wiley & Sons.

Manuele, F.A. (2012). *On the practice of safety* (4th ed.). Hoboken, NJ: John Wiley & Sons.

OSHA. (1992). *Process safety management of highly hazardous chemicals* (29 CFR 1910.119). Washington, DC: U.S. Department of Labor, Author.

14 INCIDENT INVESTIGATION
OUR METHODS ARE FLAWED

ORIGINALLY PUBLISHED OCTOBER 2014

It would be a rare exception if an outline for a safety management system did not include a requirement for incidents to be investigated and analyzed. And that is appropriate; incident investigation is a vital element within a safety management system. The comments in section E6.2 of ANSI/AIHA/ASSE Z10-2012, Standard for Occupational Health and Safety Management Systems (OHSMS) (ANSI/AIHA/ASSE, 2012, p. 25), describe the benefits that can be obtained from incident investigations:

•Incidents should be viewed as possible symptoms of problems in the OHSMS.

•Incident investigations should be used for root-cause analysis to identify system or other deficiencies for developing and implementing corrective action plans so as to avoid future incidents.

•Lessons learned from investigations are to be fed back into the planning and corrective action processes.

As Z10 proposes, organizations should learn from past experience to correct deficiencies in management systems and make modifications to avoid future incidents.

Research Results

The author has reviewed more than 1,800 incident investigation reports to assess their quality, with an emphasis on causal factors identification and corrective actions taken

(Manuele, 2013, p. 316). This revealed that an enormous gap can exist between issued investigation procedures and actual practice. On a 10-point scale, with 10 being best, an average score of 5.7 would be the best that could be given, and that could be a bit of a stretch.

These reviews confirmed that people who completed investigation reports were often biased in favor of selecting an employee's unsafe act as the causal factor and thereby did not proceed further into the investigation.

The author then conducted a five-why analysis to determine why this gap exists between issued procedures and actual practice. As the analysis proceeded, it became apparent that our model is flawed on several counts. The author's observations follow. These observations are made a priori, that is, relating to or derived by reasoning from self-evident proposition.

Why Incident Investigations May Not Identify Causal Factors

When supervisors are required to complete incident investigation reports, they are asked to write performance reviews of themselves and of those to whom they report, all the way up to the board of directors. Managers who participate in incident investigations are similarly tasked to evaluate their own performance and the results of decisions made at levels above theirs.

It is understandable that supervisors will avoid expounding on their own shortcomings in incident investigation reports. The probability is close to zero that a supervisor will write: "This incident occurred in my area of supervision and I take full responsibility for it. I overlooked X. I should have done Y. My boss did not forward the work order for repairs I sent him 3 months ago."

Self-preservation dominates, logically. This also applies to all management levels above the line supervisor. All such personnel will be averse to declaring their own shortcomings. Similarly, it is not surprising that supervisors and managers are reluctant to report deficiencies in the management systems that are the responsibility of their superiors.

With respect to operators (first-line employees) and incident causation, Reason (1990) writes:

> Rather than being the main instigator of an accident, operators tend to be the inheritors of system defects created by poor design, incorrect installation, faulty maintenance and bad management decisions. Their part is usually that of adding the final garnish to a lethal brew whose ingredients have already been long in the cooking. (p. 173)

Supervisors, one step above line employees, also work in a "lethal brew whose ingredients have already been long in the cooking." Supervisors have little or no input to the original design of operations and work systems, and are hampered with regard to making major changes to those systems. The author's practical on-site experience has shown that most supervisors do not have sufficient knowledge of hazard identification and analysis, and risk assessment to qualify them to offer recommendations for improving operating systems.

History

In safety management systems, first-line supervisors are often responsible for initiating an incident investigation report. In relatively few organizations, this responsibility is assigned to a team or an operating executive.

It is presumed that supervisors are closest to the work and that they know more about the details of what has occurred. The history on which such assignments are based can be found in three editions of Heinrich's *Industrial Accident Prevention*. Heinrich's influence continues to this day. Heinrich (1941, 1950, 1959) comments on incident investigation methods in the second, third and fourth editions of his book.

> The person who should be best qualified to find the direct and proximate facts of individual accident occurrence is the person, usually the supervisor or foreman, who is in direct charge of the injured person, The supervisor is not only best qualified but has the best opportunity as well. Moreover, he should be personally interested in events that result in the injury of workers under his control.
>
> In addition, he is the man upon whom management must rely to interpret and enforce such corrective measures as are devised to prevent other similar accidents. The supervisor or foreman, therefore, from every point of view, is the person who should find and record the major facts (proximate causes and subcauses) of accident occurrence.
>
> In addition, he and the safety engineer should cooperate in finding the proximate causes and subcauses of potential injury producing accidents. (1941, p. 111; 1950, p. 123; 1959, p. 84)

Heinrich's premise that the supervisor is best qualified to make incident investigations continues to be influential to this day, as evidenced by the following example from NSC (2009).

> Depending on the nature of the incident and other conditions, the investigation is usually made by the supervisor. This person can be assisted by a fellow worker familiar with the process, a safety professional or inspector, or an employee health professional, the joint safety and health committee, the general safety committee or a consultant from the insurance company. If the incident involves unusual or special features, consultation with a state labor department, or a federal agency, a union representative or an outside expert may be warranted. If a contractor's personnel are involved in the incident, then a contractor's representative should also be involved in the investigation.
>
> The supervisor should make an immediate report of every injury requiring medical treatment and other incidents he or she may be directed to investigate. The supervisor is

on the scene and probably knows more about the incident than anyone else. It is up to this individual, in most cases, to put into effect whatever measures can be adopted to prevent similar incidents. (p. 285)

Ferry (1981) also writes that the supervisor is closest to the action and most often is expected to initiate incident investigations. But he was one of the first writers to introduce the idea that supervisors may have disadvantages when doing so.

> The supervisor/foreman is closest to the action. The mishap takes place in his domain. As a result, he most often investigates the mishap. If it is the supervisor's duty to investigate, he has every right to expect management to prepare him for the task.
>   Yet the same reasons for having the supervisor/foreman make the investigation are also reasons he should not be involved. His reputation is on the line. There are bound to be causes uncovered that will reflect in some way on his method of operation.
>   His closeness to the situation may preclude an open and unbiased approach to the supervisor-caused elements that exist. The more thorough the investigation, the more likely he is to be implicated as contributing to the event. (p. 9)

Ferry (2009) makes similar comments about line managers and staff managers (e.g., personnel directors, purchasing agents).

> A thorough investigation often will find their functions contributed to the mishap as causal factors. When a causal factor points to their function they immediately have a point in common with the investigator. (p. 11)

In one organization whose safety director provided input for this article, the location manager leads investigations of all OSHA recordable incidents. That is terrific; senior management is involved. Many of the constraints applicable to the people who report to the manager can be overcome. But, in a sense, the manager is required to write a performance appraisal on him/herself and on the people in the reporting structure above his/her level. If contributing factors result from decisions the manager made or his/her bosses made, details about them may not be precisely recorded.

Investigation Teams

Discussions with several corporate safety professionals indicate that their organizations use a team to investigate certain incidents. Assume the team consists of supervisors who report to the same individual as the supervisor for the area in which the incident occurred. The team is expected to write a performance appraisal on the involved supervisor as well as on the person to whom all of them report, and that person's bosses.

A priori, it is not difficult to understand that supervisors would be averse to criticizing a peer and management personnel to whom they also report. The supervisor whose performance is reviewed because of an incident may someday be part of a team appraising other supervisors' performance.

At all management levels above line supervisor, it would also be normal for personnel to avoid being self-critical. Self-preservation dominates at all levels.

Safety professionals should realize that constraints similar to those applicable to a supervisor also apply, in varying degrees, to all personnel who lead or are members of investigation teams.

Nevertheless, the author found that incident investigation reports completed by teams were superior. Ferry (1981, p. 12) says, "Special investigation committees are often appointed for serious mishaps" and "their findings may also receive better acceptance when the investigation results are made public."

To the extent feasible, investigation team leaders should have good managerial and technical skills and not be associated with the area in which the incident occurred.

Chapter 7 of *Guidelines for Investigating Chemical Process Incidents* (CCPS, 2003) is titled "Building and Leading an Investigation Team." Although the word chemical appears in the book's title, the text is largely generic. The opening paragraph of Chapter 7 says:

> A thorough and accurate incident investigation depends upon the capabilities of the assigned team. Each member's technical skills, expertise and communication skills are valuable considerations when building an investigation team. This chapter describes ways to select skilled personnel to participate on incident investigation teams and recommends methods to develop their capabilities and manage the teams' resources. (p. 97)

This book is recommended as a thorough dissertation on all aspects of incident investigation. Throughout the book, competence, objectivity, capability and training are emphasized.

Training for Personnel on Incident Investigation

If personnel are to perform a function they should be given the training needed to acquire the necessary skill. Others make similar or relative comments. Ferry (1981) says, "If it is the supervisor's duty to investigate, he has every right to expect management to prepare him for the task" (p. 9).

The following citation is from *Guidelines for Investigating Chemical Process Incidents*: "High quality training for potential team members and supporting personnel helps ensure success. Three different audiences will benefit from training: site management personnel, investigation support personnel and designated investigation team members including team leaders" (CCPS, 2003, p. 105).

For each organization, several questions should be asked; the answers may differ greatly.

- How much training on hazards, risks and investigation techniques do supervisors and investigation team members receive?
- Does the training make them knowledgeable and technically qualified?
- How often is training provided?

Consideration also must be given to the time lapse between when supervisors and others attend a training session and when they complete an incident investigation report. It is generally accepted that knowledge obtained in training will not be retained without frequent use. It is unusual for team members to participate in two or three incident investigations in a year. Inadequate training may be a major problem.

What Is Being Taught: Causation Models

Dekker (2006) makes the following astute observation, worthy of consideration by all who are involved in incident investigations.

> Where you look for causes depends on how you believe accidents happen. Whether you know it or not, you apply an accident model to your analysis and understanding of failure. An accident model is a mutually agreed, and often unspoken, understanding of how accidents occur. (p. 81)

Safety professionals must understand that how they search for causal or contributing factors relates to what they have learned and their beliefs with respect to incident causation. There are many causation models in safety-related literature. Dekker (2006) describes three kinds of accident models. His models, abbreviated, are cited as examples of the many models that have been developed.

- The sequence-of-events model. This model sees accidents as a chain of events that leads up to a failure. It is also called the domino model, as one domino trips the next. *[Author's note: The domino sequence was a Heinrichean creation.]*
- The epidemiological model. This model sees accidents as related to latent failures that hide in everything from management decisions to procedures to equipment design.
- The systemic model. This model sees accidents as merging interactions between system components and processes, rather than failures within them. (p. 81)

Dekker (2006) strongly supports a systems approach to incident investigation, taking into consideration all of the relative management systems as a whole. He says:

> The systems approach focuses on the whole, not the parts. The interesting properties of systems (the ones that give rise to system accidents) can only be studied and understood when you treat them in their entirety. (p. 91)

Dekker is right: Whether persons at all levels are aware of it, they apply their own model and their understanding of how incidents occur when investigations are made. Thus, two questions need consideration:

- What have safety professionals been taught about incident causation?
- What have safety professionals been teaching people in the organizations they advise?

Answers to those questions greatly affect the quality of incident investigations. Based on the author's research (Manuele, 2011), the myths that should be dislodged from the practice of safety are:

1) Unsafe acts of workers are the principal causes of occupational incidents.

2) Reducing incident frequency will achieve an equivalent reduction in injury severity.

These myths arise from the work of Heinrich and can be found in the four editions of *Industrial Accident Prevention* (1931, 1941, 1950, 1959). Analytical evidence developed by the author indicates that these premises are not soundly based, supportable or valid.

Heinrich professed that among the direct and proximate causes of industrial incidents:

> 88% are unsafe acts of persons; 10% are unsafe mechanical or physical conditions; and 2% are unpreventable. (1931, p. 43; 1941, p. 22; 1950, p. 19; 1959, p. 22)

Heinrich advocated identifying the first proximate and most easily prevented cause in the selection of remedies for the prevention of incidents. He says:

> Selection of remedies is based on practical cause-analysis that stops at the selection of the first proximate and most easily prevented cause (such procedure is advocated in this book) and considers psychology when results are not produced by simpler analysis. (1931, p. 128; 1941; p. 269; 1950, p. 326; 1959, p. 174)

Note that the first proximate and most easily prevented cause is to be selected (88% of the time, a human error). That concept permeates Heinrich's work. It does not encompass what has been learned subsequently about the complexity of incident causation or that other causal factors may be more significant than the first proximate cause.

Many safety practitioners still operate on the belief that the 88-10-2 ratios are soundly based. As a result, they focus on correcting a worker's unsafe act as the singular causal factor for an incident rather than addressing the multiple causal factors that contribute to most incidents.

A recent example of incident causation complexity appears in the following excerpt from the report prepared by BP (2010) following the April 20, 2010, *Deepwater Horizon* explosion in the Gulf of Mexico.

> The team did not identify any single action or inaction that caused this incident. Rather, a complex and interlinked series of mechanical failures, human judgments, engineering

design, operational implementation and team interfaces came together to allow the initiation and escalation of the accident. (p. 31)

During an incident investigation, a professional search to identify causal factors such as through the five-why analysis system will likely find that the causal factors built into work systems are of greater importance than an employee's unsafe act.

The author's previous work (Manuele, 2011) covered topics such as moving the focus of preventive efforts from employee performance to improving the work system; the significance of work system and methods design; the complexity of causation; and recognizing human errors that occur at organizational levels above the worker.

Although response to that article was favorable, some communications received contained a disturbing tone. It became apparent that Heinrich's premise that 88% of occupational incidents are caused by the unsafe acts of workers is deeply embedded in the minds of some safety practitioners and those they advise. This is a huge problem. This premise was taught to students in safety science degree programs for many years and is still taught. The author received a call from one professor who said that the 2011 article gave him the leverage he needed to convince other professors that some of Heinrich's premises are not valid and should not be taught.

How big is the problem? Paraphrasing an April 2014 e-mail from the corporate safety director of one of the largest companies in the world, "We are thinking about how far to go to push Heinrich thinking out of our system. We still have some traditional safety thinkers who would squirm and voice concerns if we did that."

In May 2014, the author spoke at a session arranged by ORCHSE, a consulting organization whose members represent Fortune 500 companies. When the more than 85 attendees were asked by show of hands whether Heinrich concepts dominated their incident investigation systems, more than 60% responded affirmatively. This author believes that many of those who did not respond positively were embarrassed to do so.

At an August 2014 meeting of 121 safety personnel employed by a large manufacturing company, participants were asked: About what percentage of the incident reports at your location identify unsafe acts as the primary cause? The results follow:

% of reports	Participant responses
100%	3%
75%	33%
50%	37%
25%	12%
< 25%	15%

A total of 73% of participants indicated that for 50% to 100% of incident reports, workers' unsafe acts are identified as the primary cause. To quote the colleague who conducted this survey, "We've got work to do."

Also, note the following comments that are significant with respect to how big the problem is. For more than 35 years, E. Scott Geller has been a prominent practitioner in behavior-based safety. His current thinking is relative to the reality of causal factors and their origins. Excerpts from a recent article follow (Geller, 2014).

> A person who believes that most injuries are caused by employee behavior can be viewed as a safety bully. This belief could influence a focus on the worker rather than the culture or management systems, or many other contributing factors. As Deming warns, "Don't blame people for problems caused by the system."
>
> When safety programs are promoted on a premise such as "95% of all workplace accidents are caused by behavior," one can understand why union leaders object vehemently and justifiably to such. Claiming that behaviors cause workplace injuries and property damage places blame on the employee and dismisses management responsibility. Most worker behavior is an outcome of the work culture, the system.
>
> It is wrong to presume that behavior is a cause of an injury or property damage. Rather, behavior is one of several contributing factors, along with environmental and engineering factors, management factors, cultural factors and person-states. (pp. 41-42)

This author concludes that supervisors, management personnel above the supervisory level, investigation team members and safety practitioners who are not informed on current thinking with respect to incident causation are not qualified to identify causal and contributing factors, particularly those that derive from inadequacies in an organization's culture, operating systems and technical aspects applications, and from errors made at upper management levels. This presents a challenge for safety professionals, as well as an opportunity.

Multifactorial Aspects of Incident Causation

Most hazards-related incidents, even those that seem to be the least complex, have multiple causal factors that derive from *less than adequate* workplace and work methods design, operations management and personnel performance.

The author's reviews of incident investigation reports, mostly on serious injuries and fatalities, showed that:

- Many incidents resulting in serious injury or fatality are unique and singular events, having multiple and complex causal factors that may have organizational, technical, operational systems or cultural origins.

•Causal factors for low probability/serious consequence events are seldom represented in the analytical data on incidents that occur frequently. (Some ergonomics-related incidents are the exception.)

Those studies also showed that a significantly large share of incidents resulting in serious injuries and fatalities occurred:

•when unusual and nonroutine work is being performed;

•in nonproduction activities;

•in at-plant modification or construction operations (replacing a motor weighing 800 lb to be installed on a platform 15 ft above the floor);

•during shutdowns for repair and maintenance, and startups;

•where sources of high energy are present (electrical, steam, pneumatic, chemical);

•where upsets occur (situations going from normal to abnormal).

In every report reviewed, multiple causal factors were identified; there was an initiating event followed by a cascade of contributing factors that developed in sequence or in parallel. They related directly to deficiencies in operational management systems that should be subjects of concern when investigations are made.

Johnson (1980) writes succinctly about the multifactorial aspect of incident causation:

> Accidents are usually multifactorial and develop through relatively lengthy sequences of changes and errors. Even in a relatively well-controlled work environment, the most serious events involve numerous error and change sequences, in series and parallel. (p. 74)

Human Errors: Management Decision Making

Particular attention is given here to *Guidelines for Preventing Human Error in Process Safety* (CCPS, 1994). Although the term *process safety* appears in the book's title, the first two chapters provide an easily read primer on human error reduction.

Safety professionals should view the following highlights as generic and broadly applicable. They advise on where human errors occur, who commits them and at what level, the influence of organizational culture and where attention is needed to reduce the occurrence of human errors.

> It is readily acknowledged that human errors at the operational level are a primary contributor to the failure of systems. It is often not recognized, however, that these errors frequently arise from failures at the management, design or technical expert levels of the company. (p. xiii)

> A systems perspective is taken that views error as a natural consequence of a mismatch between human capabilities and demands, and an inappropriate organizational culture. From this perspective, the factors that directly influence error are ultimately controllable by management. (p. 3)

> Almost all major accident investigations in recent years have shown that human error was a significant causal factor at the level of design, operations, maintenance or the management process. (p. 5)
>
> One central principle presented in this book is the need to consider the organizational factors that create the preconditions for errors, as well as the immediate causes. (p. 5)

Since "failures at the management, design or technical expert levels of the company" affect the design of the workplace and the work methods (i.e., the operating system), it is logical to suggest that safety professionals encourage that incident investigations focus on improving the operating system to achieve and maintain acceptable risk levels.

Dekker's (2006) premises are pertinent to this subject. Several excerpts follow:

> Human error is not a cause of failure. Human error is the effect, or symptom, of deeper trouble. Human error is systematically connected to features of people's tools, tasks and operating systems. Human error is not the conclusion of an investigation. It is the starting point. (p. 15)
>
> Sources of error are structural, not personal. If you want to understand human error, you have to dig into the system in which people work. (p. 17)
>
> Error has its roots in the system surrounding it; connecting systematically to mechanical, programmed, paper-based, procedural, organizational and other aspects to such an extent that the contributions from system and human error begin to blur. (p. 74)
>
> The view that accidents really are the result of long-standing deficiencies that finally get activated has turned people's attention to upstream factors, away from frontline operator "errors." The aim is to find out how those "errors," too, are a systematic product of managerial actions and organizational conditions. (p. 88)
>
> The Systemic Accident Model . . . focuses on the whole [system], not [just] the parts. It does not help you much to just focus on human errors, for example, or an equipment failure, without taking into account the sociotechnical system that helped shape the conditions for people's performance and the design, testing and fielding of that equipment. (p. 90)

Reason's (1997) book, *Managing the Risks of Organizational Accidents*, is a must-read for safety professionals who want to learn about human error reduction. Reason writes about how the effects of decisions accumulate over time and become the causal factors for incidents resulting in serious injuries or substantial damage when all the circumstances necessary for the occurrence of a major event fit together. He stresses the need to focus on decision making above the worker level to prevent major incidents:

> Latent conditions, such as poor design, gaps in supervision, undetected manufacturing defects or maintenance failures, unworkable procedures, clumsy automation, shortfalls in training, less than adequate tools and equipment, may be present for many years before

they combine with local circumstances and active failures to penetrate the system's layers of defenses.

They arise from strategic and other top level decisions made by governments, regulators, manufacturers, designers and organizational managers. The impact of these decisions spreads throughout the organization, shaping a distinctive corporate culture and creating error-producing factors within the individual workplaces. (p. 10)

If the decisions made by management and others have a negative effect on an organization's culture and create error-producing factors in the workplace, focusing on reducing human errors at the worker level—the unsafe acts—will not solve the problems. Thus, the emphasis in incident investigations should be on the management system deficiencies that result in creating a negative "culture" and "error-producing factors in the workplace."

A Causation Model

Safety professionals are obligated to give advice based on a sound and studied thought process that considers the reality of the sources of hazards. The author proposes that a causation model must encompass the following premises.

•An organization's culture is the primary determiner with respect to the avoidance, elimination, reduction or control of hazards and whether acceptable risk levels are achieved and maintained.

•Management commitment or noncommitment to operational risk management is an extension of the organization's culture.

•Causal factors may derive from decisions made at the management level when policies, standards, procedures, provision of resources and the accountability system are less than adequate.

•A large majority of the problems in any operation are systemic. They derive from management decisions that establish the operating sociotechnical system—the workplace, work methods and governing social atmosphere-environment.

•A sound causation model for hazards-related incidents must consider the entirety of the socio-technical system, applying a holistic approach to both the technical and social aspects of operations. It must be understood that those aspects are interdependent and mutually inclusive.

The sociotechnical system in an organization is a derivation of its culture. The following definition of a sociotechnical system is a composite of several definitions and the author's views, based on experience.

A sociotechnical system stresses the holistic, interdependent, integrated and inseparable interrelationship between humans and machines. It fosters the shaping of both the technical and social conditions of work in such a way that both the system's output goal and the workers' needs are accommodated.

This article presents a sociotechnical causation model for hazards-related incidents. It is the author's composite and is influenced by his research and experience.

An organization's culture is established by the board of directors and senior management.

Management commitment or noncommitment to providing the controls necessary to achieve and maintain acceptable risk levels is an expression of the culture.

Causal factors may derive from shortcomings in controls when safety policies, standards, procedures, the accountability system or their implementation are inadequate with respect to the design processes and operational risk management and the inadequacies impact negatively on:

Providing resources
Risk assessments
Competency and adequacy of staff
Maintenance for system integrity
Management of change/prejob planning
Procurement, safety specifications
Risk-related systems
 •Organization of work
 •Training, motivation
 •Employee participation
 •Information, communication
 •Permits
 •Inspections
 •Incident investigation and analysis
 •Providing PPE
Third-party services
Emergency planning and management
Conformance/compliance assurance
Performance measures
Management reviews for continual improvement

Multiple causal factors derive from inadequate controls.

The incident process begins with an initiating event.

There are unwanted energy flows or exposures to harmful substances.

Multiple interacting events occur sequentially or in parallel.

Harm or damage results, or could have resulted in slightly different circumstances.

Cultural Implications That Encourage Good Incident Investigations

In one company in which management personnel are fact-based and sincere when they say that they want to know about the contributing factors for incidents, regardless of where the responsibility lies, a special investigation procedure is in place for serious injuries and fatalities.

That company's management recognized that it was difficult for leaders at all levels to complete factual investigation reports that may be self-critical. Thus, an independent facilitator serves as the investigation and discussion team leader. At least five knowledgeable people serve on the team. All team members know that a factual report is expected.

It is known that the CEO reads the reports, asks questions to ensure that the reports are complete, and sees that leaders resolve all of the recommendations made to a proper conclusion. Thus, the CEO's actions demonstrate that the organization's culture requires fact determination and continual improvement. The culture dominates and governs.

Cultural Implications That May Impede Incident Investigations

Guidelines for Preventing Human Error in Process Safety (CCPS, 1994) contains a relative and all-too-truthful paragraph related to an organization's culture:

> A company's culture can make or break even a well-designed data collection system. Essential requirements are minimal use of blame, freedom from fear of reprisals and feedback which indicates that the information being generated is being used to make changes that will be beneficial to everybody.
>
> All three factors are vital for the success of a data collection system and are all, to a certain extent, under the control of management. (p. 259)

In relation to the foregoing, the title of Whittingham's (2004) book, *The Blame Machine: Why Human Error Causes Accidents*, is particularly appropriate. According to Whittingham, his research shows that, in some organizations, a blame culture has evolved whereby the focus of investigations is on individual human error and the corrective action stops at that level. That avoids seeking data on and improving the management systems that may have enabled the human error.

What Whittingham describes is indicative of an inadequate safety culture. As an example of an aspect of a negative safety culture, consider the following real-world scenario with which this author became familiar that represents a culture of fear:

An electrocution occurred. As required in that organization, the corporate safety director visited the location to expand on the investigation. During discussion with the deceased employee's immediate supervisor, it became apparent that the supervisor knew of the design shortcomings in the lockout/tagout system, of which there were many at the location.

When asked why the design shortcomings were not recorded as causal factors in the investigation report, the supervisor responded, "Are you crazy? I would get fired if I did that. Correcting all these lockout/tagout problems will cost money and my boss doesn't want to hear about things like that."

This culture of fear arose from the system of expected performance that management created. The supervisor completed the investigation report in accord with what he believed management expected. He recorded the causal factor as "employee failed to follow the lockout/tagout procedure" and the investigation stopped there.

In such situations, corrective actions taken usually involve retraining and giving additional emphasis to the published standard operating procedure. Design shortcomings are untouched. Overcoming such a culture of fear in the process of improving incident investigation processes will require careful analysis and much persuasive diplomacy.

A Course of Action

If incident investigations are thorough and unbiased, the reality of the technical, organizational methods of operation and cultural causal factors will be revealed. If appropriate action is taken on those causal factors, significant risk reduction can be achieved. To improve incident investigation quality, safety professionals should do the necessary research and develop a plan of action.

- Safety professionals must base their practice on sound principles. They must understand the importance of and the serious need for their guidance on incident investigation to all levels of management and for investigation teams. Thus, it is suggested that safety professionals review the causation model on which their advice is based.
- A sociotechnical causation model for hazards-related incidents emphasizes the influence of an organization's culture and the shortcomings that may exist in controls when safety policies, standards, procedures and the accountability system are inadequate with respect to the design processes and operations risk management. A causation model should relate to such inadequacy of controls.
- Improving the quality of incident investigations in most organizations will require significant changes in their culture and safety professionals must understand the enormity of the task. In such an initiative, knowledge of management of change methods is necessary (Manuele, 2014).
- Valid data on the quality of incident investigations should be developed. So, an evaluation should be made of a sampling of completed investigation reports. In studies made by the author, the identification entries in incident investigation forms (e.g.,

name, department, location of the incident, shift, time, occupation, age, time in the job) received relatively high scores for thoroughness of completion.

Thus, it is suggested that the evaluation concentrate on incident descriptions, causal and contributing factor determination, and corrective actions taken. If the number of entries in an available data bank presents a manageable unit, all incident descriptions can be reviewed. As the data bank increases in size, decisions must be made about the number of incidents that practicably should be reviewed. Where the data bank is large, a safety professional may want to evaluate only incidents that result in serious injury or illness, perhaps those valued in workers' compensation claims data at $25,000 or more.

This level was selected pragmatically while working with larger companies. Safety directors decided to have the incident review process pertain to perhaps two or three or 5,000 incidents. For example, in a company in which about 5,000 workers' compensation claims are reported annually, the computer run at a $25,000 selection level provided data on 375 cases, about 7.5% of total cases. They represented more than 70% of total claims values.

•An assessment should follow of the reality of the culture in place with respect to incident investigations. This is vital. Safety professionals must understand that the culture will not be changed without support from senior management and that they must adopt a major role to achieve the necessary change.

•Other evaluations should be made to determine what is being taught about incident investigation; whether the guidance given in procedure manuals is appropriate and adequate; and whether the investigation report form assists or hinders thorough investigations.

•From the foregoing, the safety professional should draft an action plan to convince management of the value of making changes in the expected level of performance on incident investigation. One item in the action plan should propose adopting a problem-solving technique, an incident investigation technique.

The Five-Why Analysis System

The five-why analysis and problem-solving technique is easy to learn and effective; the training time and administrative requirements are not extensive. Before applying this technique, training should cover the fundamentals of hazard and risk identification and analysis. The author promotes adoption of the five-why technique rather strongly. For most organizations, achieving competence in applying the technique to investigations will be a major step forward.

The five-why concept is based on an uncomplicated premise, so it can be easily adopted in an incident investigation process, as some safety professionals have done. For the occasional complex incident, starting with the five-why system may lead to the use of event trees, fishbone diagrams or more sophisticated investigation systems.

Other incident investigation techniques exist. Highly skilled investigators may say that the five-why process is inadequate because it does not promote identification of causal factors resulting from decisions made at a senior management level. That is not so. Usually, when inquiry gets to the fourth "why," considerations are at the management levels above the supervisor and may consider decisions made by the board of directors.

Given an incident description, the investigator or the investigation team would ask "why" five times to get to the contributing causal factors and outline the necessary corrective actions. A colleague who has adopted the five-why system says that he has taught incident investigators to occasionally interject a "how could that happen?" into the discussion—an interesting innovation. A not-overly complex example of a five-why application follows.

> The written incident description says that a tool-carrying wheeled cart tipped over onto an employee while she was trying to move it. She was seriously injured.
>
> 1) Why did the cart tip over? The diameter of the casters is too small and the carts are tippy.
>
> 2) Why is the diameter of the casters too small? They were made that way in the fabrication shop.
>
> 3) Why did the fabrication shop make carts with casters that are too small? It followed the dimensions provided by engineering.
>
> 4) Why did engineering provide fabrication dimensions for casters that have been proven to be too small? Engineering did not consider the hazards and risks that would result from using small casters.
>
> 5) Why did engineering not consider those hazards and risks? It never occurred to the designer that use of the small casters would create hazardous situations. The designer had not performed risk assessments.
>
> Conclusion: I [the department manager] have made engineering aware of the design problem. In that process, an educational discussion took place with respect to the need to focus on hazards and risks in the design process. Also, engineering was asked to study the matter and has given new design parameters to fabrication: The caster diameter is to be tripled. On a high-priority basis, fabrication is to replace all casters on similar carts. A 30-day completion date for that work was set.
>
> I have also alerted supervisors to the problem in areas where carts of that design are used. They have been advised to gather all personnel who use the carts and inform them that larger casters are being placed on carts, and instruct them that until then, moving the carts is to be a two-person effort. I have asked our safety director to alert her associates at other locations of this situation and how we are handling it.

Sometimes, asking "why" as few as three times gets to the root of a problem; on other occasions, six times may be necessary. Having analyzed incident reports in which the five-why system was used, the author offers several cautions:

- Management commitment to identifying the reality of causal factors is necessary for success.
- Ensure that the first "why" is really a "why" and not a "what" or a diversionary symptom.
- Expect that repetition of five-why exercises will be necessary to get the idea across. Doing so in group meetings at several levels, but particularly at the management level, is a good idea.
- Be sure that management is prepared to act on the systemic causal factors identified as skill is developed in applying the five-why process.

A safety director who contributed material for this article says the following about his application of the five-why system.

> I have trained supervisors, shift managers, department managers and facility managers in the use of the five-why system for accident investigations. I taught them the difference between fact finding and fault finding. They understand that documenting a failure on their part does not necessarily mean that they are lousy supervisors and will help us identify system problems that we must correct. I review every investigation report. Anytime I feel they have stopped asking "why" too soon, I assist them with additional investigation to ensure that the root cause(s) are identified and appropriate corrective actions are developed and implemented.

The literature on the five-why system is not extensive because it is not complex. Two Internet resources are listed in the references for this article.

Conclusion

If incident investigations are objective and thorough, the symptoms relating to technical, organizational, methods of operation and cultural causal factors will be revealed. If appropriate action is taken on those causal factors, significant risk reduction can be achieved. But, as is established in this article, incident investigations are most often not thorough and factual.

That presents significant challenges and opportunities for safety professionals. It is incumbent on them to be well informed about incident causation. As Dekker (2006) says, "Where you look for causes depends on how you believe accidents happen. Whether you know it or not, you apply an *accident model* to your analysis and understanding of failure," (p. 81).

It is apparent that the magnitude of the need as safety professionals give advice on incident investigation and causal factor determination is huge. In most organizations, a major culture change will be necessary to significantly improve the quality of incident investigations, a change that can be achieved only with management support over time.

Assume that a safety professional decides to take action to improve the quality of incident investigation. It is proposed that the following comments about incident investigation, as excerpted from the Report of the *Columbia* Accident Investigation Board (NASA, 2003), be kept in mind as a base for reflection throughout the endeavor.

> Many accident investigations do not go far enough. They identify the technical cause of the accident, and then connect it to a variant of "operator error." But this is seldom the entire issue.
>
> When the determinations of the causal chain are limited to the technical flaw and individual failure, typically the actions taken to prevent a similar event in the future are also limited: fix the technical problem and replace or retrain the individual responsible. Putting these corrections in place leads to another mistake—the belief that the problem is solved.
>
> Too often, accident investigations blame a failure only on the last step in a complex process, when a more comprehensive understanding of that process could reveal that earlier steps might be equally or even more culpable. In this Board's opinion, unless the technical, organizational, and cultural recommendations made in this report are implemented, little will have been accomplished to lessen the chance that another accident will follow. (p. 177)

Paraphrasing, for emphasis: If the cultural, technical, organizational and methods of operation causal factors are not identified, analyzed and resolved, little will be done to prevent recurrence of similar incidents. ■

References

ANSI/AIHA/ASSE. (2012). *American national standard for occupational health and safety management systems*. (ANSI/AIHA/ASSE Z10-2012). Des Plaines, IL: ASSE.

BP. (2010, Sept. 8). *Deepwater Horizon accident investigation report*. Retrieved from http://cdm16064.contentdm.oclc.org/cdm/ref/collection/p266901coll4/id/2966

Center for Chemical Process Safety (CCPS). (1994). *Guidelines for preventing human error in process safety*. New York, NY: American Institute of Chemical Engineers (AIChE).

CCPS. (2003). *Guidelines for investigating chemical process incidents* (2nd ed.). New York, NY: AIChE.

Dekker, S. (2006). *The field guide to understanding human error*. Burlington, VT: Ashgate Publishing Co.

Ferry, T.S. (1981). *Modern accident investigation and analysis: An executive guide*. Hoboken, NJ: John Wiley & Sons.

Geller, E.S. (2014, Jan.) Are you a safety bully? Recognizing management methods that can do more harm than good. *Professional Safety, 59*(1), 39-44.

Heinrich, H.W. (1931). *Industrial accident prevention*. New York, NY: McGraw-Hill Book Co.

Heinrich, H.W. (1941). *Industrial accident prevention* (2nd ed.). New York, NY: McGraw-Hill Book Co.

Heinrich, H.W. (1950). *Industrial accident prevention* (3rd ed.). New York, NY: McGraw-Hill Book Co.

Heinrich, H.W. (1959). *Industrial accident prevention* (4th ed.). New York, NY: McGraw-Hill Book Co.

Johnson, W.G. (1980). *MORT safety assurance systems.* New York, NY: Marcel Dekker.

Manuele, F.A. (2011, Oct.). Reviewing Heinrich: Dislodging two myths from the practice of safety. *Professional Safety, 56*(10), 52-61.

Manuele, F.A. (2013). *On the practice of safety* (4th ed.). Hoboken, NJ: John Wiley & Sons.

Manuele, F.A. (2014). *Advanced safety management: Focusing on Z10 and serious injury prevention* (2nd ed.). Hoboken, NJ: John Wiley & Sons.

Mapwright Pty Ltd. The 5 whys method. Essendon, Australia: Author. Retrieved from www.mapwright.com.au/5-whys-method.html

MoreSteam.com. 5-why analysis. Powell, OH: Author. Retrieved from www.moresteam.com/toolbox/5-why-analysis.cfm

NASA. (2003, Aug.). *Columbia* accident investigation report (Vol. 1, Chapter 7). Washington, DC: Author. Retrieved from www.nasa.gov/columbia/home/CAIB_Vol1.html

National Safety Council (NSC). (2009). *Accident prevention manual for business and industry: Administration and programs* (13th ed.). Itasca, IL: Author.

Reason, J. (1990). *Human error.* New York, NY: Cambridge University Press.

Reason, J. (1997). *Managing the risks of organizational accidents.* Burlington, VT: Ashgate Publishing Co.

Whittingham, R.B. (2004). *The blame machine: Why human error causes accidents.* Burlington, MA: Elsevier Butterworth-Heinemann.

15 CULTURE CHANGE AGENT
THE OVERARCHING ROLE OF OSH PROFESSIONALS

ORIGINALLY PUBLISHED DECEMBER 2015

All safety professionals should view all hazardous situations as indicators of inadequacies in the safety management processes that relate to the existence of these situations. Assume that management takes corrective action to eliminate every hazardous situation identified. Safety professionals should realize that relatively little will be gained if no effort is made to eliminate the management system deficiencies. Eliminating those deficiencies will require changes in an organization's culture.

This idea, in a sense, extends the goal of the planning requirements established in Section 4 of ANSI/ASSE Z10-2012, Occupational Health and Safety Management Systems, which states, "The planning process goal is to identify and prioritize occupational health and safety management system issues (defined as hazards, risks, management system deficiencies and opportunities for improvement) and to establish objectives" (p. 9).

Overcoming management systems deficiencies occurs only by modifying the way things get done—that is, only if an organization's culture is changed with respect to its system of expected performance. Thus, the safety professional's overarching role is that of a culture change agent. To substantiate that premise, this article:
- defines overarching, systems, processes, culture and culture change agent;
- provides examples of situations in which safety professionals did not recommend the necessary systems and culture changes;

- reviews job descriptions;
- comments on a safety culture within an organization's overall culture and how the advice given by safety professionals affects the culture;
- recognizes the difficulties when the safety culture is negative;
- provides resources with respect to safety professionals as culture change agents;
- proposes that the proposition made here be tested against the requirements of ANSI/ASSE Z10.

Definitions

Let's begin with a review of several key terms.

- **Overarching:** A composite definition is:

> Encompassing everything; embracing all else; including or influencing every part of something. This premise, that the safety professional's overarching role is that of a culture change agent, applies universally to all who give advice on improving operational risk management systems.

- **Systems and processes:** These terms are discussed because systems and processes must be modified to achieve culture change. These terms are commonly used for the elements of an operational risk management system. For example, in ANSI/ASSE Z10, the words *process* and *system* appear 120 times in the first numbered 29 pages. From the many definitions available, those presented are from the online Business Dictionary (www.businessdictionary.com/definition/system.htm).

a) Process: Sequence of interdependent and linked procedures which, at every stage, consume one or more resources (employee time, energy, machines, money) to convert inputs (data, material, parts) into outputs. These outputs then serve as inputs for the next stage until a known goal or end result is reached.

b) System: An organized, purposeful structure that consists of interrelated and interdependent elements (components, entities, factors, members, parts). These elements continually influence one another (directly or indirectly) to maintain their activity and the existence of the system, in order to achieve the goal of the system.

One can argue that the terms *process* and *system* are synonyms. When a process or a system is modified and that modification is successful, a culture change is achieved.

- **Culture:** Many definitions of *safety culture* are available. Most, if not all, imply that harmony exists with respect to safety at all levels of employment. Composite definitions follow, representing definitions typically found in the literature.

a) Safety culture is defined as entrenched attitudes and the shared values, beliefs, assumptions and norms that may govern organizational decision making.

b) Safety culture reflects attitudes, beliefs, perceptions and values that employees at all levels share.

c) Safety culture refers to ingrained attitudes and opinions that a particular group of people share with respect to risk and safety.

The literature does not, however, provide a description of a safety culture that recognizes that management decisions made over time may result in the existence of a multitude of unacceptable risks within an operation. The cumulative result in such situations is a negative culture, one in which harmony does not exist; in which shared values or common beliefs are few; and in which group attitudes about safety are negative. In some organizations, employers may believe that certain operational risks are acceptable, while employees may have other views and conclude that those risks are unacceptable. Thus, employees may not share the views and beliefs that management holds with respect to safety and operational risk levels.

An organization's safety culture, which is a subset of its overall culture, derives from decisions made at the governing entity level (e.g., board of directors, group of owners) and at the senior management level that result in acceptable or unacceptable operational risk levels. Outcomes of those decisions could be positive or negative. Safety is culture driven, and management establishes the culture. An organization's culture is translated into a system of expected performance that defines the staff's beliefs with respect to what management wants done.

Although an organization may issue commendable safety policies, manuals and operating procedures, the staff's perception of what is expected of them and the performance for which they will be measured—the system of expected performance—may differ from what is written.

In reality, what management does may differ from what management says. What management does defines an organization's culture and its commitment or noncommitment to safety. Employee perceptions of management's position on safety are, in effect, employees' reality. These perceptions, realistic or unrealistic, are their truths.

Safety, defined as being free from unacceptable risks, will improve only if the culture changes, that is, only if the system of expected performance undergoes major changes.

•**Change agent:** Definitions of this term are numerous as well. The following composite fits well with the safety professional's position.

> A change agent is a person who serves as a catalyst to bring about organizational change. A change agent assesses the present, is controllably dissatisfied with it, contemplates a future that should be, and takes action to achieve the culture changes necessary to achieve the desired future.

Overlooking Necessary Systems and Culture Changes

A study aiming to improve actions on serious injury and fatality prevention concluded that safety professionals should identify risk situations that could be precursors to serious injuries and fatalities and recommend corrective action to eliminate those precursors. It was not suggested that deficiencies in management systems relative to the existence of these precursors be identified. However, it is highly probable that if a company makes no changes in the relative management systems and decision making, then those systems will continue to create such precursors.

An organization decided to initiate a prevention through design system involving all operations personnel. Employees were educated in hazard identification and analysis, and risk assessment. Numerous situations were identified for which risks should be reduced, and the design revisions for the workplace and the work methods were commendable. The safety professionals involved were asked how these revisions were communicated to design personnel. They indicated that no such communication had occurred, that no attempts were made to change the original design system. Not sharing information about the design measures taken meant that designers would continue to produce designs that included the risky work situations.

Job Descriptions for Safety Managers

For simplification, the term *safety manager* was selected as a caption for the many examples of job descriptions for safety professionals. These descriptions may encompass environmental risks and OSH risks. A few also include product safety, waste management or fire protection. They refer to functions (e.g., develops, performs, assists, analyzes, implements), but rarely delineate an overall purpose. Several job descriptions contain the following statement: "Models and promotes an organizational culture that fosters safe practices through effective leadership." (For an example, visit http://bit.ly/1PAZcM8.)

However, none of the job descriptions reviewed state that the safety professional's overarching role is that of culture change agent. As the OSH field continues its progress toward becoming recognized as a profession, that role should be understood, and that awareness should influence how individual practitioners promote the practice of safety.

Safety Professionals and Safety Culture

What is the safety professional's role with respect to an organization's safety culture? Assume that safety is a core value in an organization and that senior management is determined to achieve and maintain acceptable risk levels in all operations. Usually, the safety professionals in such organizations are well qualified, they have stature, and the advice they give is well received and seriously considered.

Even in such organizations, change, favorable or unfavorable, is a constant. The safety management systems in place will continually develop information indicating that certain safety-related processes can be improved. Then, acting from a sound professional base, the advice given by safety professionals in their role as culture change agents is welcomed, mostly. In this role, they:

- Perform diligent data gathering and analysis to identify process shortcomings.
- Propose and arrange for the performance of hazard identification and analyses, and risk assessments.
- Give advice on prioritizing risks.
- Recommend actions that management should take for improvement.

However, not all organizations have superior safety management systems in place. In these settings, being a culture change agent is more difficult, particularly if senior management believes that all is well and that changes are unnecessary. The safety professional's operating base in such situations remains the same as the bulleted list just offered, but the skill level required to be a successful culture change agent can be exceptionally demanding. Therefore, patience is required. Satisfaction may derive principally from small steps forward. But the goal remains the same: Try to positively influence the safety culture toward achieving acceptable risk levels.

Now, assume that the organization's culture has always been negative or is drifting into a negative state. Safety professionals must recognize and discuss this concept of drift, particularly since significant expense reductions in recent years have caused some companies to drift into a negative state with respect to safety. As Rasmussen (1997) writes:

> The scale of industrial installations is steadily increasing with a corresponding potential for large-scale accidents. Companies today live in a very aggressive and competitive environment which will focus the incentives of decision makers on short-term financial and survival criteria rather than long-term criteria concerning welfare, safety and environmental impact. (p. 186)

The word *drift* has been attached to Rasmussen's premise. For example, Dekker (2011) references Rasmussen's work:

> Drift occurs in small steps. This can be seen as decrementalism, where continuous adaptation around goal conflicts and uncertainty produces small, step-wise normalizations of what was previously judged as deviant or seen as violating some safety constraint. (p. 15)

When a safety professional senses drift occurring due to decisions that result in violating some safety constraint, or more likely several safety constraints, as a culture change agent, the safety professional must attempt to deter or slow the pace of drift. Using the same diligent data-gathering and analysis methods to identify process shortcomings and risk prioritization, a safety professional can counsel management on the facts and the pace of the drift toward danger. The goal is to make management aware that the organization is putting in place elements that increase the potential for a large-scale incident and to encourage management to slow or stop the pace of deterioration in processes. Assessing and prioritizing risks and emphasizing the growing potential for the occurrence of low-probability, severe-consequence events acquire greater importance.

At sites where safety culture has drifted into a negative state or has always been negative, safety professionals likely have limited resources. Thus, their communications with management should contain information on safety-related decisions that should be made, on a priority basis, so that these limited resources can be applied to achieve the greatest good. Again, this requires priority setting and focusing in particular on preventing low-probability, serious-consequence events.

A safety professional will not likely easily achieve success in such situations. However, the probability of success will be enhanced if s/he is viewed as an integral member of the business team. That will result from giving well-supported, substantiated, and convincing technical and managerial risk management advice that is perceived as serving the organization's interests.

Thus, to be successful culture change agents, safety professionals must operate within the business framework of the organizations to which they give counsel. Thus, safety professionals should seek to obtain additional knowledge and skills that will bolster their qualifications to do so:
- business management basics;
- financial analysis tools, such as cost/benefit analysis;
- language of finance, which is the language of management;
- budgeting process;
- impact of adequate or inadequate cash flow;
- elements that may influence executive decision making.

Relevant Resources

Spigener and Groover (2008) discuss the safety professional's emerging role as change agent:

Staying relevant as an organization changes [means] learning how to leverage your knowledge, skills and experience in new ways.... If you are a technical expert in [EHS], the good news is that you already have the skills and knowledge to contribute to safety strategy. The hard part will be gaining fluency in organizational change management.

Spigener and Groover (2008) also identify core competencies of change agents. They:

- are forever inquisitive and never-ending learners;
- advance performance by identifying what ought to be, deciding how to get there and influencing decision makers to adopt their ideas;
- do not leave their expertise behind;
- leverage their knowledge and experience to develop strategies to positively influence actions that result in higher performance levels;
- recognize that to be influential in achieving change, they must acquire change management skills;
- become aware of the culture in place and learn how to manage within it to effect change;
- recognize the effect of management decisions and actions on the culture;
- find ways to tactfully inform management when they believe that management decisions and actions may have negative results.

Simon (1999) opens his culture change chapter in *Safety Through Design* (Christensen & Manuele, 1999) as follows: "A full explanation of what culture change is and is not, who is involved, why it is necessary and can achieve world-class safety through design, and how to make it happen is provided in this chapter" (p. 37). Although the chapter focuses on safety through design, it is generic with respect to what is required to achieve a culture change.

Swuste and Arnoldy (2003) examine the challenge of becoming an agent of organizational change. Let's review their article's abstract.

> There is a great need for health and safety advisers/managers to act as agents of change, both in respect to the technology of the company and the design of its workplaces, and in the organization of the company health and safety management system. This article reports on the development of training to meet these increasing needs. The postgraduate masters course "Management of Safety, Health and Environment" of the Delft University of Technology has now introduced a course-module of 1 week, addressing the issue of the learning organization and the specific role of the safety adviser/manager.
>
> The course-module starts from the assumption that for a health and safety adviser/manager his/her personal effectiveness and ability to influence and stimulate others are qualities as important to a company as the quality of a safety and health management system. This paper will describe the development in the role of the safety adviser/manager and the mainstream thinking on change management and training. The consequence for the content and program features of the course-module is presented as well as the results of the evaluation of its effectiveness.

For emphasis, let's consider again this sentence:

> The course-module starts from the assumption that for a health and safety adviser/manager his/her personal effectiveness and ability to influence and stimulate others are qualities as important to a company as the quality of a safety and health management system.

Kello (2005) also examines this topic as shown in the following excerpts.

> So what is the proper role of the safety professional in the total safety culture, to which many organizations today aspire? It is definitely not the same old technical expert role, even with a broader bandwidth. It is fundamentally, qualitatively different in its approach.
>
> In the field of organization development [OD], OD practitioners have been referred to as "change agents" from the very beginnings of the discipline. My central thesis is that, whether they normally think of it in these terms or not, to be truly effective in the flexible, team-based high-performance organization, safety professionals must perform as change agents, too. In my view, much like organization development consultants, safety professionals encourage and help people make constructive behavior change, to do things differently, to challenge longstanding habits and to get out of their "comfort zone." Further, and also like the OD consultant, they are almost always more of an influencer than a director.

Kello (2005) adds, "Modern safety professionals are agents for positive change in their organizations. They are trying to build deep working relationships that allow them to effect constructive change through influence, even when the client system may not want to change."

A U.S. Department of Defense slide presentation pertaining to patient safety identifies and discusses the eight steps of change [citing Kotter (1996) as the source of these steps] in order to help viewers:

- Describe the actions required to set the stage for organizational change.
- Identify ways to empower team members to change.
- Discuss what is involved in creating a new culture.
- Begin planning for the change in the organization.

If safety professionals presented this slide series as they attempt to influence others on how to achieve organizational change, they would be serving as culture change agents. Readers are encouraged to review this slide series for its informative value.

Why Culture Change Initiatives Fail

Kotter's (1996) book *Leading Change* is a foundational work. The thought pattern presented is largely based on what Kotter learned from practical applications. As he reports, many change initiatives fail. As culture change agents, safety professionals should be well informed on how change initiatives succeed and fail, and on how success and failure are measured. The following references address failed initiatives:

- "Five Reasons Why Leaders Fail to Create Successful Change" (Eikenberry, 2005).
- "Seven Reasons Why Organizational Change Fails" (Brazier, 2007).
- "Why Change Efforts Typically Fail: 15 Predictable Reasons & Situations to Avoid" (posted at http://bit.ly/20L3Dck; based on Blanchard, 2009).
- "Leading Change: Why Transformation Efforts Fail" (Kotter, 2007).

Experience shows that change initiatives fail for various reasons. The first reason is the most important.

1) The culture embedded in place and how to work within it is largely ignored.

2) Leadership and commitment necessary at sufficiently high levels to achieve the change may not in reality exist because the change agent has not invested the time necessary to achieve the required commitment.

3) Decision makers are not seriously enthusiastic about the change proposed because the supporting data are shallow and unconvincing.

4) The importance of becoming aware of the power structure and determining how to work within it has not been sufficiently recognized.

5) Team building, which is vital to success, has been inadequate.

6) Preparing for the typical resistance to change at all levels comes up short.

7) Communication to all personnel levels that would be affected by the change is less thorough than needed.

8) Management personnel who are assigned responsibility for the change may not be held accountable for progress by those to whom they report and, in time, the urgency and importance for the change diminishes.

9) People assume that a process or system has changed without determining that it has. Some refer to this as declaring victory too soon. Some advise that one should not claim success in culture modification until at least 1 year has passed. Too often, supervisors and operators revert to previous methods.

10) Change agents are not sufficiently aware that achieving a culture change may take a long time.

A Basic Guide

Environmental Management Systems: An Implementation Guide for Small and Medium-Sized Organizations (NSF International, 2001) is largely devoted to environmental management, yet many parts of the downloadable guide are generic and, thus, basic to almost all change initiatives. Readers are encouraged to add this publication to their professional resource library. A few excerpts follow.

Objectives and Targets

Objectives and targets help an organization translate purpose into action. These environmental goals should be factored into your strategic plans. This can facilitate the integration of environmental management with your organization's other management processes.

You determine what objectives and targets are appropriate for your organization. These goals can be applied organization-wide or to individual units, departments or functions—depending on where the implementing actions will be needed.

In setting objectives, keep in mind your significant environmental aspects, applicable legal and other requirements, the views of interested parties, your technological options, and financial, operational, and other organizational considerations.

There are no "standard" environmental objectives that make sense for all organizations. Your objectives and targets should reflect what your organization does, how well it is performing and what it wants to achieve.

Hints

•Setting objectives and targets should involve people in the relevant functional area(s). These people should be well positioned to establish, plan for, and achieve these goals. Involving people at all levels helps to build commitment.

•Get top management buy-in for your objectives. This should help to ensure that adequate resources are applied and that the objectives are integrated with other organizational goals.

•In communicating objectives to employees, try to link the objectives to the actual environmental improvements being sought. This should give people something tangible to work towards.

•Measureable objectives should be consistent with your overall mission and plan and the key commitments established in your policy (pollution prevention, continual improvement and compliance). Targets should be sufficiently clear to answer the question, "Did we achieve our objectives?"

•Be flexible in your objectives. Define a desired result, then let the people responsible determine how to achieve the result.

•Objectives can be established to maintain current levels of performance as well as to improve performance. For some environmental aspects, you might have both maintenance and improvement objectives.

•Communicate your progress in achieving objectives and targets across the organization. Consider a regular report on this progress at staff meetings.

•To obtain the views of interested parties, consider holding an open house or establishing a focus group with people in the community. These activities can have other payoffs as well.

•How many objectives and targets should an organization have? Various EMS implementation projects for small- and medium-sized organizations indicate that it is best to start with a limited number of objectives (say, three to five) and then expand the list over time. Keep your objectives simple initially, gain some early successes and then build on them.

•Make sure your objectives and targets are realistic. Determine how you will measure progress towards achieving them. (p. 29)

A Test of the Premise

As noted, the content and purposes of the planning section of ANSI/ASSE Z10-2012 support the premise that a safety professional's overarching role is that of culture change agent. If management system deficiencies exist, safety professionals must propose changes to eliminate those deficiencies. If those changes are successful, the culture will be changed. Attempting to achieve culture changes is a foundational practice of safety.

As stated at ANSI/ASSE Z10, E1.1, "This standard provides basic requirements for occupational health and safety management systems, rather than detailed specifications" (p. 1). Z10 is a management system standard and it clearly indicates that management must have certain systems and processes in place to conform with the standard. Correcting system and process shortcomings will require changes in the system of expected performance. If the revisions are permanent, the culture will be changed.

Following is a portion of the table of contents from ANSI/ASSE Z10.

ANSI/ASSE Z10: Partial Table of Contents
3.0 Management Leadership and Employee Participation
   3.1 Management Leadership
      3.1.1 Occupational Health and Safety Management System
      3.1.2 Policy
      3.1.3 Responsibility and Authority
   3.2 Employee Participation
4.0 Planning
   4.1 Review Process
   4.2 Assessment and Prioritization
   4.3 Objectives
   4.4 Implementation Plans and Allocation of Resources
5.0 Implementation and Operation
   5.1 OHSMS Operational Elements
      5.1.1 Risk Assessment
      5.1.2 Hierarchy of Controls
      5.1.3 Design Review and Management of Change
      5.1.4 Procurement
      5.1.5 Contractors
      5.1.6 Emergency Preparedness
   5.2 Education, Training, Awareness and Competence
   5.3 Communication
   5.4 Document and Record Control Process
6.0 Evaluation and Corrective Action
   6.1 Monitoring, Measurement and Assessment
   6.2 Incident Investigation
   6.3 Audits
   6.4 Corrective and Preventive Actions
   6.5 Feedback to the Planning Process

7.0 Management Review
　7.1 Management Review Process
　7.2 Management Review Outcomes and Follow Up

Readers are asked to review each section and subsection listed to try to locate exceptions to the premise that overcoming management system deficiencies will require changes in the system of expected performance and, thus, an organization's culture. It is difficult to conceive of a situation in which a hazard, a risk or a management system deficiency exists for which a change in a process and the system of expected performance are not a remedy.

If the process change is successful, the organization will achieve acceptable risk levels. However, remember this caution: One must examine revisions in a process and, thereby, a culture change, over time. Such examination will confirm whether personnel have reverted to previous practices after what seems to be a success in the short term; ensure that the modification is delivering the expected outcomes; and verify that no unintended consequences are created that increase risk.

Conclusion

One can make a sound case to support the premise that safety professionals' overarching role is that of culture change agent. Thus, to enhance their capability and effectiveness, safety professionals should:

•Recognize the validity of this premise.

•Focus on the premises set forth in ANSI/ASSE Z10 that "the planning process goal is to identify and prioritize occupational health and safety systems issues (defined as hazards, risks, management system deficiencies and opportunities for improvement)."

•Be aware that hazardous situations are indicators of inadequacies in the safety management processes that relate to these situations' existence and that corrective actions must eliminate those deficiencies in order to be deemed adequate.

•Become familiar with change management principles (e.g., resources cited in this article).

•Develop a champion at the senior management level by presenting the results of analyses of incidents and the deficiencies in the relative management systems that should be eliminated for effective operational risk management.

•Recognize that achieving culture changes may take a long time.

While serving as the managing director for a safety and fire protection consultancy, the author recognized that giving advice to clients was the only product the firm sold. Advice given was based on the staff's superior technical and managerial knowledge and skill. Success was determined by whether clients believed that this advice provided value relative to the fees paid.

Think about it. Are not most safety professionals primarily providers of advice to achieve change? Safety professionals are most often in staff positions and their role is to provide advice to decision makers on hazards, risks and deficiencies in management systems so that changes can be enacted to achieve acceptable risk levels and culture change.

Many organizational change initiatives fail because the prevailing culture is ignored. Achieving permanent change is difficult. Therefore, safety professionals must understand an organization's existing culture; acknowledge how deeply certain practices are embedded within processes; determine who presumes to have ownership of them; and embrace the need to plan and communicate to achieve change. Safety professionals must be perceived as members of the management team. Change methods they adopt should align with the procedures with which the management team is comfortable in order to avoid major conflict. Safety professionals have always been culture change agents. By recognizing the reality of this role, they could become more effective in counseling management and influencing decisions. ■

References

ANSI/ASSE. (2012). *Occupational health and safety management systems (ANSI/ASSE Z10-2012)*. Park Ridge, IL: ASSE.

Blanchard, K. (2009). *Leading at a higher level*. Upper Saddle River, NJ: Pearson/Prentice Hall.

Brazier, A. (2007). Seven reasons why organizational change fails. Retrieved from http://andybrazier.blogspot.com/2007/04/seven-reasons-why-organisational-change.html

Christensen, W. & Manuele, F. (Eds.) (1999). *Safety through design*. Itasca, IL: NSC.

Dekker, S. (2011). *Drift into failure*. Burlington, VT: Ashgate Publishing Co.

Eikenberry, K. (2005). Five reasons why leaders fail to create successful change. Retrieved from http://ezinearticles.com/?Five-Reasons-Why-Leaders-Fail-to-Create-Successful-Change&id=26088

Kello, J. (2005). Changing the safety culture: Safety professionals as change agents. *The International Journal of Knowledge, Culture and Change Management*, 6(4), 151-156.

Kotter, J.P. (1996). *Leading change*. Boston, MA: Harvard Business Review Press.

Kotter, J.P. (2007). Leading change: Why transformation efforts fail. *Harvard Business Review*. Retrieved from http://www.srfmr.org/uploads/teaching_resource/1396040272-211c50caa4635c59b/Leading%20Change.pdf

NSF International Strategic Registrations Ltd. (2001) *Environmental management systems: An implementation guide for small- and medium-sized organizations* (2nd ed.). Retrieved from www2.epa.gov/ems/environmental-man-agement-systems-implementation-guide-small-and-medium-sized-organizations

Rasmussen, J. (1997). Risk management in a dynamic society: A modeling problem. *Safety Science, 27*(2/3), 183-213.

Simon, S. (1999). Achieving the necessary culture change. In W. Christensen and F. Manuele (Eds.), *Safety through design*. Itasca, IL: NSC.

Spigener, J. & Groover, D. (2008). Staying relevant: The emerging role of the safety professional: Becoming a change agent. Retrieved from http://bstsolutions.com/en/knowledge-resource/269-staying-relevant-the-emerging-role-of-the-safety-professional-part-2-becoming-a-change-agent

Swuste, P. & Arnoldy, F. (2003). The safety adviser/manager as agent of organizational change: A new challenge to expert training. *Safety Science, 41*(1), 15-27.

U.S. Department of Defense. Change management: How to achieve a culture of safety. Retrieved from www.ahrq.gov/sites/default/files/wysiwyg/professionals/education/curriculum-tools/teamstepps/longtermcare/module8/igltcchangemgmt.pdf

16 Root-Causal Factors
Uncovering the Hows and Whys of Incidents

Originally published May 2016

This article began after reading some thought-provoking comments about incident causation by authors Erik Hollnagel and Sidney Dekker. Hollnagel is the author of *Barriers and Accident Prevention* (2004), and Dekker wrote *The Field Guide to Understanding Human Error* (2006). OSH professionals should read the writings of both. Consider some of their commentary:

1) One can describe and understand an incident in several ways, and the cause-effect assumption is perhaps the least attractive option (Hollnagel, p. 26).

2) The tendency to look for causes rather than explanations is often reinforced by the methods used for incident analysis. The most obvious example of that is the principle of root-cause analysis (Hollnagel, p. 26).

3) Root cause is a meaningless concept (Hollnagel, p. 28).

4) There is no root cause (Dekker, p. 77).

5) What you call root cause is simply the place where you stop looking any further (Dekker, p. 77).

6) Where you look for causes depends on how you believe incidents happen. Whether you know it or not, you apply an accident model to your analysis and understanding of failure (Dekker, p. 81).

The positions Hollnagel and Dekker take are educational and promote introspection. Safety professionals should analyze these positions for their possible effects on the practice of safety. The excerpts in this article are intentionally presented like book reviews. This was done to illustrate the breadth of what these noteworthy authors have written about how incidents happen, incident causes, root causes and incident analysis.

In addition to reviewing and commenting on statements made by Hollnagel and Dekker, this article presents a concept that OSH professionals can practically apply to determine root-causal factors. Identifying incident causal/contributing factors has long been a basic element in safety management systems. Simply stated, the purpose of an incident investigation is to learn from history and to make improvements to overcome the management system deficiencies noted in investigation reports.

Hollnagel on Causation

Hollnagel's (2004) first two chapters, which span 58 pages, are titled "Accidents and Causes" and "Thinking About Accidents." He says that when investigating incidents, applying a cause-and-effect approach is the least attractive of all options. His premise is that causal/contributing factors do not occur sequentially in complex events (p. 26); for incidents that are not complex, a cause-and-effect method may be sufficient. Thus, a safety practitioner could determine its applicability by assessing the simplicity or complexity of a given organization's hazard/risk environment.

Hollnagel (2004) implies that investigators should seek the hows and whys of incidents, expressed in narrative descriptions, rather than seeking causes. He pleads for an understanding of the difference between explanations of the hows and whys an incident occurs and seeking causes (p. 26). He is particularly opposed to seeking root-causal factors (p. 26). However, this reasoning is difficult to follow because if an incident's hows and whys are determined, they are more than likely the causal factors.

Hollnagel (2004) recommends using prescribed causal factor models during an incident investigation, but says that because of their structure and content these models may interfere with or limit the process of determining the how and why of an event. That is an acceptable premise. Hollnagel says root-cause analysis is an example of a less-than-adequate incident investigation method, calling the concept deceptive. In Hollnagel's view, the method's name implies that the product of an investigation will be *the* root cause (p. 27). It is exceptionally important to note that Hollnagel refers to root-cause analysis in the singular. When he describes the root-causal concept as meaningless, he refers to attempts to find the one and only root cause of a problem (p. 28). Look for more about singularity later in this article.

Hollnagel cites philosophers (e.g., Nietzche, p. 25; Hume, p. 31) and what they have written about how difficult it is to determine the reality of causes. Generally, these philosophers' stance on cause is that there can be no uncaused cause; everything that exists has causes for its existence; and no matter how deeply one delves to identify causes, reasons will emerge for the existence of the identified causes and, thus, the attempt will be never ending.

Therefore, if an investigation stops because investigators become comfortable with their causal factor determinations, philosophers would argue that inquiry could continue. So, assume that as an investigation proceeds, management systems deficiencies are identified. While desirable to identify the reasons for their existence (e.g., senior management decisions), doing so when the investigations are performed by internal personnel would be an exception. Practicably, the investigation process stops when those involved conclude that their inquiry has reached its cultural and organizational limits.

Although Hollnagel (2004) repeats his view that it is difficult to define what a cause is, he offers the following plausible definition: "A cause can be defined as the identification, after the fact, of a limited set of aspects of the situation that are seen as the necessary and sufficient conditions for the observed effect(s) to have occurred" (p. 34).

This definition is consistent with applied and practical incident investigation processes, and it is one that safety professionals can confidently support. Hollnagel (2004) acknowledges that identifying incident causes is instructive and valuable in determining corrective actions (p. 32).

Hollnagel (2004) ends his first chapter by stating that determining cause is a "relative and pragmatic" venture, but that doing so is not "scientific." This is another logical premise. Some decisions made during an incident investigation result from what people say and, thereby, may be subjective and not necessarily reflective of good science. Nevertheless, in support of determining the correct causal factors, Hollnagel (2004) says, "The value of finding the correct cause or explanation is that it becomes possible to do something constructively to prevent future accidents" (p. 35).

As noted, the purpose of an incident investigation is to learn from history and to make improvements to overcome any management system deficiencies identified. That closely fits what Hollnagel says about the value of finding correct causes or explanations.

Dekker on Causation

Dekker's (2006) *The Field Guide to Understanding Human Error* is particularly thought provoking because of the positions he takes on human error and how those views relate to his comments on incident investigation. Consider these excerpts.

- Human error is not a cause of failure. Human error is the effect, or symptom, of deeper trouble. Human error is systematically connected to features of people's tools, tasks and operating systems. Human error is not the conclusion of an investigation. It is the starting point. (p. 15)

- Sources of error are structural, not personal. If you want to understand human error, you have to dig into the system in which people work. (p. 17)

- Error has its roots in the system surrounding it; connecting systematically to mechanical, programmed, paper-based, procedural, organizational and other aspects to such an extent that the contributions from system and human error begin to blur. (p. 74)

- The view that accidents really are the result of long-standing deficiencies that finally get activated has turned people's attention to upstream factors, away from frontline operator "errors." The aim is to find out how those "errors" too, are a systematic product of managerial actions and organizational conditions. (p. 88)

Dekker's (2006) ninth chapter is titled "Cause Is Something You Construct." Some of his views mirror those expressed by Hollnagel (2004). For example, Dekker (2006) writes, "The reality is that there is no such thing as the cause, or primary cause or root cause. Cause is something we construct, not find. And how we construct depends on the accident model we believe in" (p. 73).

Note that the wording is singular as is Hollnagel's (2004) text. Dekker (2006) makes it clear that he is opposed to seeking a singular cause or singular root cause when investigating incidents.

Leveson (2011) also questions the concept of root cause. She implores incident investigators to use an incident model that promotes a "broad view" of an occurrence.

> An accident model should encourage a broad view of accident mechanisms that expands the investigation beyond the proximate events. A narrow focus on operator actions, physical component failures, and technology may lead to ignoring some of the most important factors in terms of preventing future accidents. The whole concept of "root cause" needs to be reconsidered. (p. 33)

In a sense, Leveson (2011) supports the views expressed by Hollnagel (2004) and Dekker (2006) when she says that debating whether one causal factor is the root-causal factor among other causal factors wastes time and can be nonproductive (p. 56).

Dekker (2006) asserts that incident investigators do not find causes, but rather surmises that they construct them. That is, what an investigator identifies as causes is influenced by his/her assumptions about how incidents occur (p. 73).

This latter point, that causal factor determination is influenced by the investigator's beliefs about how incidents occur, is an important truism that should prompt extensive introspection and self-analysis by safety practitioners. For example, suppose an investigator believes that unsafe acts of workers are the principal causes of occupational incidents. That belief would have a significant effect as that investigator participates in determining causal factors.

Now assume that safety practitioners have taught management that unsafe acts are the principal causes of occupational incidents (which has occurred). Management personnel would then have the same understanding of how incidents occur and the investigation system would be grossly inadequate.

Synonyms for construct are *build*, *make*, *fabricate* and *create*. Dekker's (2006) statement is a reflection of his overall position that investigators do not in reality discover what causes incidents. In the author's experience, the determination of causal factors is often dismally shallow and is neither found nor created. Unfortunately, in a huge proportion of investigations, the causal factor determination stops with identifying an employee's unsafe act.

Again, the writings of Hollnagel (2004) and Dekker (2006) are substantially similar. Dekker writes:

> There is no "root" cause. What you call "root cause" is simply the place where you stop looking any further. You see that factor as necessary for the mishap to happen. Nothing else would have needed to go wrong; otherwise you would also have to label those things as "causes." (p. 77)

Dekker (2006) implies that an investigation system is deficient if it requires the identification of a single root cause. Dekker and Hollnagel (2004) also agree that there is no such a thing as a root cause and that it is folly to try to establish a root cause for an occurrence that has several contributing causes. The author's research supports this. "Many incidents resulting in serious injury are unique and singular events, having multiple and complex causal factors that may have organizational, technical, operating systems or cultural origins" (Manuele, 2014a, p. 62).

Dekker (2006) says that it would be better if investigators wrote about "explanations rather than causes" (p. 78). Note the term *explanations*. It appears several times in Hollnagel's (2004) book.

Like Hollnagel (2004), Dekker (2006) recognizes that some organizations require investigators to identify root-causal factors when their process is complete. His recommendation on what to do when such requirements exist is comparable to Hollnagel's as well. Dekker suggests that the investigator write a narrative, an explanation of the incident's how and why, that includes the issues and events that the investigator believes to be important, with designations of probable causes being subordinate (p. 78). As noted, the author believes that the how and why of an incident are likely the causal factors.

Hollnagel and Dekker on Incident Models

Hollnagel (2004) writes extensively about the need for an organization to select an incident model to serve as a base for incident investigation, communication and corrective action. Dekker's (2006) comments are not extensive, but they mirror Hollnagel's (2004) thoughts on three types of models.

Sequential Models

Sequential models view incidents as a result of a sequence of events that occur in a specific order. Heinrich's domino sequence was the first such model. Like many others, Hollnagel (2004) observes that sequential models are inadequate when incidents result from multiple causal factors that may contribute to an incident in parallel rather than sequentially (p. 47). He also continues to be nonsupportive of seeking a root cause (p. 51).

In discussing these models, Hollnagel (2004) comments further on the possible deterrent effects of applying the "stop rule" to determine when an investigation is complete. International standard IEC 62740:2015, Root Cause Analysis, includes a reference about the stop rule: "The 'stopping rule' is the point at which action can be defined or additional proof of cause is no longer necessary for the purpose of the analysis" (IEC, 2015, p. 15).

Hollnagel raises a valid point. Investigations typically stop too soon, thereby avoiding recognition of the management systems deficiencies that could be important among the causal factors. However, even if an investigation meets the stop rule requirements of the international standard, it is likely that certain philosophers (and Hollnagel and Dekker) would find the causal factor determination insufficient, believing that no matter where the investigation stops, it could have gone further.

Epidemiological Models

An epidemiological model would show an incident as deriving from a blend of causal factors, some active and others built up over time and existing in combination at the time the incident occurred. Hollnagel (2004) says that an epidemiological model considers performance deviations; deviations from safe practice; environmental conditions (technical and social aspects); the absence or ineffectiveness of preventive barriers; and latent hazardous conditions or practices that have accumulated over time (p. 54).

Systems Models

Application of a systems model requires one to consider individual systems as interrelated and inseparable parts of a whole. Hollnagel (2004) recognizes the value of a form of linear plotting (perhaps several linear plottings such as for a fishbone diagram) that may be causally related (p. 59). He writes, "Events can still be ordered post hoc either temporally or in terms of causal relations. But in the systemic model each event may be preceded by several events (temporally or causally), as well as be followed by several events" (p. 59).

Hollnagel (2006) recommends the use of a systems model (as does the author). Applying a systems model requires macro thinking rather than micro thinking. Using micro thinking, one would, for example, hold that unsafe acts are the principal causal factors for occupational incidents and stop an investigation once a worker's unsafe act is identified. Taking a macro view to determine causal factors requires thinking large about systems as integral and inseparable parts of a whole and their interrelationships. It also requires seeking management systems deficiencies, some of which could derive from an organization's cultural, technical and social aspects.

Hollnagel's (2004) fifth chapter, "A Systemic Accident Model," includes a depiction of his functional resonance as a systemic accident model (FRAM) (Hollnagel & Goteman, 2015, p. 171). FRAM is based on four principles: 1) the equivalence of failures and successes; 2) the central role of approximate adjustments; 3) the reality of emergence; and 4) functional resonance as a complement to causality. Hollnagel's thinking is new and interesting, yet because of the model's complexity, some safety practitioners will find it difficult to accept. In organizations with deeply embedded investigation systems, obtaining acceptance of FRAM will require a concentrated, multiyear effort to achieve the culture change necessary.

Dekker's (2006) 10th chapter is titled "What Is Your Accident Model?" He recognizes that a model helps one determine what is to be sought, yet he also says that a model may be restraining. He extends previous statements about how a person's understanding and beliefs about how incidents happen influences the thoroughness of an investigation. His observation has substantial weight and safety professionals should seriously consider it.

> Where you look for causes depends on how you believe accidents happen. Whether you know it or not, you apply an accident model to your analysis and understanding of failure. An accident model is a mutually agreed, and often unspoken, understanding of how accidents occur. (p. 81)

Dekker (2006) uses the same three model groups as Hollnagel (sequential, epidemiological, systems), and his comments on these models mirror Hollnagel's. Dekker also favors the use of systemic models. One can make a convincing case that root-causal factors (plural) can be most effectively identified by focusing on the whole of the sociotechnical aspects of operations as an interacting system (Manuele, 2014a, Chapter 5).

How many incident models currently exist? Safety Institute of Australia (2012) issued a document that highlights 32 such models. This document is available as a free download (http://bit.ly/1XluQPI).

Comments on Root Causes and Causal Factors

Entering the phrase "root causes of accidents" into a search engine will return an abundance of resources. Let's focus on selections from two publications.

Petersen (1998) comments on the concept of multiple causation and how an investigation into root causes should identify shortcomings in management systems. Note that Petersen writes in the plural.

> Multiple causation asks what are some of the contributing factors surrounding an incident? If we deal only at the symptomatic level, we end up removing symptoms but allowing root causes to remain to cause another accident or some other type of operational error.
>
> Root causes often relate to the management system. They may be due to management's policies and procedures, supervision and its effectiveness, or training. Root causes are those which would effect permanent results when corrected. They are those weaknesses which not only affect the single accident being investigated, but also might affect many other future accidents and operational problems. (p. 11)

Center for Chemical Process Safety (CCPS, 2003) publishes guidelines for investigating chemical process incidents, which include information on structured approaches to determining root causes. CCPS's definition of *root cause* begins in the singular, then recognizes that incidents usually have more than one root cause. "Root cause: A fundamental, underlying, system-related reason why an accident occurred that identifies a correctable failure(s) in management systems. There is typically more than one root cause for every process safety incident" (p. 179).

This definition is particularly noteworthy. It states that investigators are to seek "system-related reason(s)" and "failure(s) in management systems." Great emphasis must be given to examining the operating system that management creates. Reason (1990) appeals for recognition of the significance of system defects when discussing latent errors and system disasters. As they participate in or give counsel on incident investigation, safety professionals should seriously consider Reason's insight:

> Rather than being the main instigators of an accident, operators tend to be the inheritors of system defects created by poor design, incorrect installation, faulty maintenance and bad management decisions. Their part is usually that of adding the final garnish to a lethal brew whose ingredients have already been long in the cooking. (p. 173)

An International Standard

As noted, an international standard exists for root-cause analysis (RCA). Consider this abstract about the standard (IEC, 2015):

> IEC 62740:2015 describes the basic principles of root-cause analysis (RCA) and specifies the steps that a process for RCA should include. This standard identifies a number of attributes for RCA techniques which assist with the selection of an appropriate technique. It describes each RCA technique and its relative strengths and weaknesses.
>
> RCA is used to analyze the root causes of focus events with both positive and negative outcomes, but it is most commonly used for the analysis of failures and incidents. Causes for such events can be varied in nature, including design processes and techniques, organizational characteristics, human aspects and external events.
>
> RCA can be used for investigating the causes of nonconformances in quality (and other) management systems as well as for failure analysis, for example in maintenance or equipment testing. RCA is used to analyze focus events that have occurred, therefore this standard only covers a posteriori analyses.
>
> The intent of this standard is to describe a process for performing RCA and to explain the techniques for identifying root causes. These techniques are not designed to assign responsibility or liability, which is outside the scope of this standard.

Given how standards-development committees work, it is understandable that the number of definitions for causal factors listed in the standard's definition category is excessive. They are repetitive and overlap, and use of all of them in an investigation system would promote valueless discussion and inefficiency. Nevertheless, all are listed here. Note that the definition given for *root cause* is in the singular but that the explanation transitions into the plural.

- **Cause:** Circumstance or set of circumstances that leads to failure or success.
- **Causal factor:** Condition, action, event or state that was necessary or contributed to the occurrence of the focus event.
- **Contributory factor:** Condition, action, event or state regarded as secondary, according to the occurrence of the focus event.
- **Focus event:** Event that is to be explained causally.
- **Root cause:** Causal factor with no predecessor that is relevant for the purpose of the analysis.
  1) A focus event normally has more than one root cause.
  2) In some languages, the term root cause refers to the combination of causal factors that have no causal predecessor (a cut set of causal factors).
- **Root-cause analysis:** Systematic process for identifying the causes of a focus event.
- **Stopping rule:** Reasoned and explicit means for determining when a causal factor is defined as being a root cause. (p. 9)

This standard was reaffirmed in 2015. Its existence indicates that RCA is alive and well in many places around the world.

The Five-Why Problem-Solving Technique

Some consider the five-why problem-solving technique to be overly simplistic. However, research has revealed that the quality of causal factor determination as shown in incident investigation reports is often poor, even in large organizations (Manuele, 2014b). Thus, because of the observed status quo and what can be practicably attained, the author strongly recommends the five-why system as an initial undertaking to improve incident investigation quality.

The five-why system is easy to learn and apply to improve incident investigation quality. Case in point: One large company determined that if the five-why system were promoted and the company were able to give itself a B+ grade on investigation quality in 2 years, the company would have taken huge steps forward.

The technique consists simply of asking why five times consecutively. It is important that the first step identify a why, and not a what or a who. Sometimes, asking why only four times is sufficient. Occasionally, the process requires six or seven inquiries. Furthermore, in some situations, interjecting an occasional what or a how may move the inquiry forward. Occasionally, applying the technique to interrelated systems identifies actions that several operational entities should take to resolve a problem.

Safety professionals should select the categories of injuries or losses for which they would propose the use of the five-why system. Because of the time and expense involved and the limited benefits to be obtained, the system should not be used for minor incidents such as a paper cut or a scratch from an improperly set staple. However, it can produce beneficial results when applied to major incidents.

What is a major incident? Following is a composite list of the major incident categories established in various guidelines. Safety professionals can select from or add to this list as they develop a definition suitable to the organizations to which they give counsel.

- OSHA-reportable incident;
- hospitalization of an employee for more than 3 days;
- incident resulting in injury to three or more employees;
- a fatality;
- incident that did not result in harm or damage, but could have had serious consequences under other circumstances;
- incident resulting in property damage in excess of $10,000;
- product loss valued in excess of $10,000;
- environmental incident that was reported to a governmental authority;
- incident that required building or job site evacuation;
- incident that required emergency shutdown of operations;

Chapter 16: Root-Causal Factors

- incident that could generate public interest;
- extraordinary or unusual incident creating a crisis or significant emergency;
- incident that will provide a lesson learned for other locations.

When using this technique, it is best to select a review team with suitable experience and knowledge. The team leader should have solid managerial and technical skills. To the extent feasible, the team leader should not be associated with the area in which the incident occurred.

CCPS (2003) provides helpful guidance on building and leading an investigation team.

> A thorough and accurate incident investigation depends upon the capabilities of the assigned team. Each member's technical skills, expertise and communication skills are valuable considerations when building an investigation team. This chapter describes ways to select skilled personnel to participate on incident investigation teams and recommends methods to develop their capabilities and manage the teams' resources. (p. 97)

Four examples of five-why application follow.

Example 1: Design Flaw

The written incident description reports that a tool-carrying wheeled cart tipped over while an employee was trying to move it. She was seriously injured.

1) Why did the cart tip over? The diameter of the casters is too small and the carts are tippy. This has happened several times but there was no injury.

2) Why weren't the previous incidents reported? We didn't recognize the extent of the hazard and that a serious injury could occur when a cart tipped over.

3) Why is the diameter of the casters too small? They were made that way in the fabrication shop.

4) Why did the fabrication shop make carts with casters that are too small? They followed the dimensions given to them by engineering.

5) Why did engineering give fabrication dimensions for casters that have been proven to be too small? Engineering did not consider the hazards and risks that would result from using small casters.

6) Why did engineering not consider those hazards and risks? It never occurred to the designers that use of small casters would create hazardous situations.

Causal/contributing factors: Hazard was not recognized by operations personnel; design of the casters resulted in hazardous situations.

Conclusion: I [the department manager] have made engineering aware of the design problem. In that meeting, I emphasized the need to focus on hazards and risks in the design process. Also, engineering was asked to study the matter and has given new design parameters to fabrication: the caster diameter is to be tripled. On a high-priority basis, fabrication is to replace all casters on similar carts. We set a 30-day completion date for that work.

I also alerted supervisors to the problem in areas where carts of that design are used. They have been advised that when deciding to report or not report an incident that did not result in injury, they are to be extra cautious. And, they have been advised to gather all personnel who use the carts and instruct them that larger casters are to be placed on tool carts and that, until that is done, moving the carts is to be a two-person effort. I have asked our safety director to alert her associates at other locations of this situation and how we are handling it.

Example 2: Materials Variation

Operations personnel report concern over injury potential resulting from conditions that develop when a metal-forming machine stops because the overload trip actuates. This scenario offers an example of how the five-why technique can be used to resolve hazard/risk situations before an incident occurs. The safety director met with the supervisor who is directly responsible for the work.

1) Why are you concerned? The electrical overload trip actuates very often when we use this machine. It gets risky when it stops in midcycle and the work needed to clear the partially formed metal adds risks that employees think are more than they should have to bear. Occasionally, that's okay; often is too much.

2) Why does the overload trip actuate? This is a new problem for us. We rarely had the overload trip actuate. It started after a new order for metal was received. We are told that the purchasing department thought that it got a good deal from a metals distributor, but what was delivered did not meet our specifications. This metal is not as malleable and workable, and the metal former struggles in the forming process. So, the overload trip actuates. Maintenance is furious with us because we have to call them so often.

3) Why can't the amperage for the overload trip be increased for this batch of metal? Our engineers say they don't want greater power fed into this machine.

4) Why do you have to call on maintenance so often? The rule here is that no overload trip is to be reset without a review of why it tripped and clearance from maintenance.

5) Why haven't you recommended to your operating manager that he meet with the engineer and maintenance manager to decide how to resolve the overload trip problem for this batch of metal? That's not easy for me to do at my level. But, it would be good if you could find a way to get that done.

Possible causal/contributing factors: Overexertion; machine actuating when cleaning the partially formed metal; fall potential; partially formed metal presents hazard in the handling process.

Resolution of this risk-related problem involves the purchasing department (with respect to future purchases), operations, engineering and maintenance. Often, risk-reduction actions require participation by several interrelated and integrated functions. It is also clear that the supervisor does not feel free to discuss a hazardous situation with his boss.

Example 3: Electrical Safety

The technician fixing a broken machine did not turn off the electric power and was electrocuted. Initially, the causal factor was recorded as the unsafe act of the technician who did not follow the lockout/tagout procedure.

1) Why would the technician take such a shortcut? He was under considerable pressure from production to get the machine back in operation.

2) Why would production put that much pressure on him? This machine is vital to the overall process and production was lagging. Some production people were idle and doing nothing.

3) Why did the technician not take the few steps needed to follow the lockout/tagout procedure? We checked and found that the lockout/tagout station was not nearby.

4) How far away was the station? More than 300 ft.

5) Why was the station located so far away? That's the way the system was designed. We have a lot of situations where the lockout/tagout station is not nearby. They have been discussed but it was decided not to move them.

6) How could this situation be resolved? Senior management is upset about this fatality. So, engineering and maintenance are preparing a list of all lockout/tagout situations that need attention. We have been told that the work will get done.

Causal factors/contributing factors: distance to the lockout/tagout station made it inconvenient to go there; management did not recognize the risks of not having lockout/tagout stations nearby; production's pressure to get the repairs finished.

Example 4: Poor Maintenance

A machine operator slipped, fell and broke a hip. Oil on the floor. Cleaned the floor. (These three sentences are exactly what was stated in the report for the incident description, the causal factors and the corrective action. Further inquiry followed.)

1) Why was there oil on the floor? A gasket leaked.

2) Why did the gasket leak? Bearings are worn on this machine and when it is stressed, it vibrates a lot. The vibration loosened the joint.

3) Why is the machine stressed? When production is at full pace, which is often, this machine just barely meets the demand.

4) Why haven't the bearings been replaced? We sent two work orders with no response.

5) Why hasn't maintenance responded? We have been through two cost reductions and maintenance is short staffed. They prioritize work orders and ours have not reached sufficient priority status.

6) Why hasn't the machine been replaced with one that can handle production at full pace? That has been discussed at department meetings, but we haven't been able to get approval.

7) How could this situation be resolved? Management has been alerted about this issue and the potential for similar problems with other machines that are not being properly maintained. We hope management adds maintenance staff. Our department head has submitted a request to acquire a machine with larger capacity.

Causal/contributing factors: leaking gasket; worn bearings; maintenance staff's inability to respond to work orders on a timely basis; and operating a machine beyond its capacity.

Some incident investigation experts criticize the five-why technique because it may not address management system deficiencies. If the discussions focus on why and how questions, such deficiencies emerge around the fourth inquiry. The results would be close to cause-and-effect relationships. Safety professionals must understand that when this technique is applied to complex operations, the results may indicate that a more sophisticated causal factor determination system is needed. For the five-why system to be effective, management must establish that it wants to be informed about the reality of causal factors.

Conclusion

When investigating incidents, safety professionals must consider the amount of time available and organizational limits. If the investigation process recognizes deficiencies in management systems, the place at which investigators stop may be at the realistic organizational boundary.

Consider the internally prepared report on the *Deepwater Horizon* explosion (BP, 2010). The executive summary contains the following terms: "the causal chain of events"; "possible contributing factors"; and "caused this accident." Here's an excerpt from the executive summary.

> The team did not identify any single action or inaction that caused this accident. Rather, a complex and interlinked series of mechanical failures, human judgments, engineering design, operational implementation and team interfaces came together to allow the initiation and escalation of the accident. Multiple companies, work teams and circumstances were involved over time. (p. 11)

The second sentence in the excerpt contains five subjects indicating management system deficiencies. If an investigation system requires in-depth inquiry to include identification of causal factors such as mechanical failures, human judgment, engineering design, operational implementation and team interfaces, the system stops at a high management level and causal/contributing factors are found, not created.

Some may suggest that investigations should continue further to determine how each management system deficiency came to be. While that would be nice to do, investigators could say that getting as far into management system deficiencies as the BP team did may be as far as internally employed investigators can practically go due to cultural and organizational limitations. The BP report would receive a superior rating in relation to the quality of causal factors identified compared to the more than 1,800 investigation reports the author has reviewed.

If an investigation process determines how and why an incident occurs and identifies the deficiencies in management systems, the contributing factors (the root-causal factors if one elects to so name them) are identified.

Improving incident investigation quality is much more important than the terminology an organization adopts. The ultimate goal is to achieve superior investigations. If an incident's why and how are cited in investigation reports, the investigators will have determined the root-causal factors and arrived at an appropriate stopping point. If the system in place reveals multiple causal/contributing factors and it works, what the factors are called is of less significance. Safety professionals cannot let semantics get in the way of accomplishment.

If what an organization has in place is effective, it is best to stick with it. Although incident investigations may not achieve absolute certainty in determining root-causal factors, having recognized that uncertainty, safety professionals can give advice that can be practically applied with respect to root-causal factors and the management system improvements required. ■

References
ANSI/ASSE. (2012). *Occupational health and safety management systems* (ANSI/ASSE Z10-2012). Des Plaines, IL: ASSE.

BP. (2010, Sept. 8). *Deepwater Horizon*: Accident investigation report—Executive summary. Retrieved from www.bp.com/content/dam/bp/pdf/sustainability/issue-reports/Deepwater_Horizon_Accident_Investigation_Report_Executive_summary.pdf

Center for Chemical Process Safety (CCPS). (2003). *Guidelines for investigating chemical process incidents* (2nd ed.). New York, NY: Author.

Dekker, S. (2006). *The field guide to understanding human error*. Burlington, VT: Ashgate Publishing.

Hollnagel, E. (2004). *Barriers and accident prevention*. Burlington, VT: Ashgate Publishing.

Hollnagel, E. & Goteman, O. (2015, March 18). The functional resonance accident model (FRAM). Retrieved from www.skybrary.aero/bookshelf/books/403.pdf

International Electrotechnical Commission (IEC). (2015). Root-cause analysis (IEC 62740:2015). Geneva, Switzerland: Author. Retrieved from https://webstore.iec.ch/publication/21810

Leveson, N.G. (2011). *Engineering a safer world: Systems thinking applied to safety*. Cambridge, MA: The MIT Press.

Manuele, F.A. (2013). *On the practice of safety* (4th ed.). Hoboken, NJ: John Wiley & Sons.

Manuele, F.A. (2014a). *Advanced safety management: Focusing on Z10 and serious injury prevention* (2nd ed.). Hoboken, NJ: John Wiley & Sons.

Manuele, F.A. (2014b, Oct.). Incident investigation: Our methods are flawed. *Professional Safety, 59*(10), 34-43.

Petersen, D. (1998). *Safety management: A human approach* (2nd ed.). Des Plaines, IL: ASSE.

Reason, J. (1990). *Human error*. Cambridge, U.K.: Cambridge University Press.

Safety Institute of Australia Ltd. (2012). Models of causation: Safety. Retrieved from www.ohsbok.org.au/wp-content/uploads/2013/12/32-Models-of-causation-Safety.pdf?4ddbe2

Whittingham, R.B. (2004). *The blame machine: Why human error causes accidents*. Burlington, MA: Elsevier Butterworth-Heinemann.

17 HIGHLY UNUSUAL
CSB'S COMMENTS SIGNAL LONG-TERM EFFECTS ON THE PRACTICE OF SAFETY

ORIGINALLY PUBLISHED APRIL 2017

CSB's report on the *Deepwater Horizon* incident contains several unusual comments pertaining to operations risk management that may have long-term effects on the practice of safety. CSB is a well-regarded governmental agency. This article calls attention of OSH professionals to these comments.

The agency's final report on the explosion and fire that occurred April 20, 2010, at the Macondo *Deepwater Horizon* rig in the Gulf of Mexico was issued April 12, 2016. The incident resulted in 11 fatalities, 17 injuries and extensive environmental damage. CSB's comments may be signals indicating that, over time, organizations should revise their accountability levels and the content of their operations risk management systems that aim to protect people, property and the environment.

The report's executive summary sets forth CSB's responsibility: "CSB is an independent federal agency charged with investigating industrial chemical accidents. Its mission is to independently investigate significant chemical incidents and hazards and to effectively advocate for implementing its recommendations to protect workers, the public and the environment" (CSB, 2016a, p. 11).

While CSB investigates and reports on chemical incidents, safety professionals should consider as generic the sections of its report on the Macondo event addressed by this article.

According to the executive summary, "BP was the main operator/lease holder responsible for the well design and Transocean was the drilling contractor that owned and operated the *Deepwater Horizon* drilling rig" (CSB, 2016a, p. 6).

The executive summary of CSB's (2016a) report on the Macondo incident explains the report's four volumes:

> Volume 1 recounts a summary of events leading up to the Macondo explosions and fire. (p. 9)

> Volume 2 explores several technical findings related to the functioning of BOP [blow out preventer], a subsea system that was intended to mitigate or prevent a loss of well control. (p. 9)

> Volume 3 explores human and organizational factors associated with the incident, including aspects of the decision making by the well operations crew. (p. 10)

> Volume 4 delves into the role of the safety regulator in overseeing offshore oil and gas activities. (p. 10)

Comments in this article pertain to select sections in the executive summary, and volumes 3 and 4. These excerpts show the unusual nature of CSB's language.

Boards of Directors

The author is not aware of another governmental incident investigation report, or any other incident investigation report, that prominently implies that inadequacies at an organization's board-of-director level may contribute to the occurrence of a major incident. Volume 3, which explores human and organizational factors, makes many references to what a board of directors should do. These references are more precisely stated in the executive summary.

The following excerpts sufficiently demonstrate that CSB (2016a) believes that boards of directors have a responsibility to provide more extensive stewardship than they have in the past, and to hold the executive staff accountable with respect to the avoidance of major incidents.

> Corporate board of directors' oversight, shareholder activism, and U.S. Securities and Exchange Commission reporting requirements have the potential to influence an organization's focus on major accident risk. (p. 9)

> Post-Macondo industry and regulatory gaps in managing safety-critical elements, human factors, process safety indicators, corporate governance, workforce engagement, and major accident risk management and oversight need to be filled. (p. 9)

> [Volume 3] also addresses strategies for ensuring boards of directors remain focused on potential major accident events by examining corporate governance good practice. (p. 10)

To paraphrase, as they fulfill their corporate governance responsibilities, boards of directors are to focus on major incident potential and provide oversight to avoid the possible occurrence of such incidents. Few organizations operate this way. Yet, the logic of CSB's proposal is not easily refuted.

The author reviewed his own body of work to determine whether his writing states as precisely as CSB's report what the agency says with respect to corporate responsibility. It does not. The nearest the author's work comes is in *Advanced Safety Management: Focusing on Z10 and Serious Injury Prevention* (Manuele, 2014), which describes "a sociotechnical model for an operational risk management system":

The board of directors and senior management establish a culture for continual improvement that requires defining, achieving and maintaining acceptable risk levels in all operations.

Management leadership, commitment, involvement and the accountability system establish that the performance level to be achieved is in accord with the culture established by the board.

To achieve acceptable risk levels, management establishes policies, standards, procedures and processes with respect to:

Providing adequate resources
Risk assessment, prioritization and management
  •Applying a hierarchy of controls
Prevention through design
  •Inherently safer design
  •Resiliency, reliability and maintainability
Competency and adequacy of staff
  •Capability—skill levels
  •Sufficiency in numbers
Maintenance for system integrity
Management of change/prejob planning
Procurement—safety specifications
Risk-related systems
  •Organization of work
  •Training—motivation
  •Employee participation
  •Information—communication
  •Permits
  •Inspections
  •Incident investigation and analysis
  •Providing PPE
Third-party services
  •Relationships with suppliers
  •Safety of contractors—on premises
Emergency planning and management
Compliance/compliance assurance reviews

> **Performance measurement: Evaluations are made and reports are prepared for management review to support continual improvement and to ensure that acceptable risk levels are maintained.**

With respect to a board of directors and senior management, the model says:

> The board of directors and senior management establish a culture for continual improvement that requires defining, achieving and maintaining acceptable risk levels in all operations.
> Management leadership, commitment, involvement and the accountability system establish that the performance level to be achieved is in accord with the culture established by the board to achieve acceptable risk levels. (p. 110)

Note that the preceding excerpt does not reference a board of directors as having specifically defined responsibility to, as it provides governance, focus on major incident potential and provide oversight to avoid the possible occurrence of such incidents. The author does not know of a resource that does so.

What could result from the positions taken by CSB? Transitions in the practice of safety occur slowly. But, look ahead 5 to 10 years and CSB's influence may be substantively felt. OSH professionals should be attentive to opportunities to provide the necessary advice that may affect the guidance a board of directors is expected to provide as well as the organizational culture created by the board and senior management.

Organizational Culture

The section titled "Culture for Safety: Focus and Response" in Volume 3 of the CSB (2016b) report addresses organizational culture. CSB places extensive emphasis on the influence that an organization's culture can have on avoiding major incidents. That a government entity gives such importance to culture is most unusual and surprising. Three excerpts from Volume 3 speak to this point.

> "A strong safety culture cannot eliminate all accidents, especially in technologically complex and dynamic industries such as deepwater drilling. There is always a risk that an accident will happen. Strong safety cultures can reduce the likelihood of accidents and the severity of accidents should they occur." (Sutcliffe, 2011, as cited in CSB, 2016b, p. 236)

> A culture that truly promotes safety extends beyond workers' perceptions, espoused values and documented policies. . . . A culture for safety is characterized not only by goals, policies and procedures, but by the *company's commitment to them and what it actually does* [Emphasis added]. (p. 240)

> Thus, a company's most senior leadership, starting at the board of directors, plays the pivotal role in influencing a culture that robustly promotes process safety. Cases show that actual practices repeated by a group over time, when enforced and verified by an authoritative entity, can lead to a culture change (Hopkins, p. 1). Institutional actions

offer deep insight into a corporate culture: "critical controls to prevent a major incident are just another way of describing important organizational practices" (Wilkinson, 2016, as cited in CSB, 2016b, p. 242).

How significant to read that CSB agrees that "a company's most senior leadership, starting at the board of directors, plays the pivotal role in influencing a culture" and that this subject is of such importance as to be included in its report.

OSH professionals should agree that management creates and owns the culture, and that an organization's culture is a prominent determinant with respect to the occurrence of incidents. Rarely can an organization's culture be developed from the bottom up.

The author has written extensively about this subject. The following excerpt from Chapter 7 of *On the Practice of Safety*, titled "Superior Safety Performance: A Reflection of an Organization's Culture," is an example.

Culture Defined and Its Significance

If an entity wants to achieve superior safety results, safety must become a core value within the organization's culture. Safety is culture-driven. Where safety is a core value within a company, senior management is personally and visibly involved and holds employees at all levels accountable for results.

Senior executives display by what they do that safety is a subject to be taken very seriously. An organization's culture determines the level of safety to be obtained. What the board of directors or senior management decides is acceptable for the avoidance, elimination and control of hazards is a reflection of its culture. Management attains, as a derivation of its culture, the hazards-related incident experience it establishes as tolerable. For personnel in an organization, "tolerable" is their interpretation of what management does. (Manuele, 2013, p. 126)

Another example is the following excerpt from Chapter 8 of *Advanced Safety Management: Focusing on Z10 and Serious Injury Prevention*, titled "Management Leadership and Employee Participation: Section 3.0 of Z10."

Safety is culture driven, and management establishes the culture. Management owns the culture. An organization's safety culture is represented by the reality of application of its goals, performance measures, and sense of responsibility to its employees, to its customers, and to its community—all of which are translated into *a system of expected performance*. Over the long term, the injury experience attained is a direct reflection of an organization's safety culture. Strong emphasis is given to the phrase *a system of expected performance* because it defines what the staff believes that management, in reality, wants done. Although organizations may issue safety policies, manuals and standard operating procedures, the staff's perception of what is expected of them and the performance for which they will be measured—its *system of expected performance*—may differ from what is written. (Manuele, 2014, p. 142)

An organization will achieve major improvements in safety only if a culture change takes place—only if significant changes occur in an organization's *system of expected performance*.

Risk Assessment, Risk Reduction and ALARP

Volume 3 (CSB, 2016b) consists of comments made by a prestigious government agency that promote making risk assessments and establishing realistic risk reduction goals. Note particularly, the agency says that the outcome of those activities should achieve risk levels that are as low as reasonably practicable (ALARP). Although the term ALARP often appears in safety-related literature and in some standards, it is most unusual for a federal agency to adopt the concept implied by ALARP as a risk level to be achieved—and sufficient—in risk management.

It should be understood that, while conducting a risk assessment in itself does not guarantee that the risks will be managed, the act of conducting a risk assessment provides an organization the opportunity to identify and control those risks. Consider the following excerpts from Volume 3 (CSB, 2016b).

> Companies need an effective, and realistic, risk reduction goal because they cannot eliminate every risk completely—absolute safety is not possible. The question then becomes, when are efforts to reduce the level of residual risk sufficient? This challenge led to reducing risk to a level as low as is reasonably practicable, or ALARP, an important concept to explore in risk reduction practices. (p. 170)

> ALARP is also defined as "efforts to reduce risk [that are] continued until the incremental sacrifice (in terms of cost, time, effort or other expenditure of resources) is grossly disproportionate to the incremental risk reduction achieved" (CCPS, 2007, p. xxxvii). In practice, these efforts by the company are twofold. First, they are the initial identification and implementation of physical, operational/human, and organizational safety barriers to reduce the risk of a major accident as determined by a hazard analysis. Second, they are adherence to safety management systems intended to ensure strong barriers throughout the lifetime of an operation. The success of these systems hinges on the risk management approach and corporate oversight of that approach to create a strong and supportive culture. (p. 171)

> While an initial effort to address risk levels is necessary, the efforts should be continual and in response to various factors such as new technology developments, updated industry standards or lessons learned from an incident. (p. 171)

What does all this mean? Safety professionals should understand that absolute safety is not attainable. No matter how extensive the consideration of hazards and risks in the design and operation phases, residual risk will always exist. The residual risk should be as low as is reasonably practicable and acceptable, and risk must be continuously assessed as situations change. This is a strong statement, especially for a government entity to imply that reducing risks to a level as low as is reasonably practicable is tolerable and acceptable.

ALARP seems to be an adaptation from ALARA, which is as low as reasonably achievable. Use of the ALARA concept as a guideline originated in the atomic energy field.

As defined in Title 10, Section 20.1003, of the *Code of Federal Regulations* (10 CFR 20.1003), ALARA is an acronym for "as low as (is) reasonably achievable," which means making every reasonable effort to maintain exposures to ionizing radiation as far below the dose limits as practical, consistent with the purpose for which the licensed activity is undertaken, taking into account the state of technology, the economics of improvements in relation to state of technology, the economics of improvements in relation to benefits to the public health and safety, and other societal and socioeconomic considerations, and in relation to utilization of nuclear energy and licensed materials in the public interest. (NRC, 2017)

The implication that decision makers are to "[make] every reasonable effort to maintain exposures to ionizing radiation as far below the dose limits as practical" provides conceptual guidance when striving to achieve acceptable risk levels in all classes of operations.

ALARP has become the more frequently used term for operations outside the atomic energy arena. Concepts embodied in the terms ALARA and ALARP apply to the design of products, facilities, equipment, work systems and methods, and environmental controls, as well as in operations risk management.

In the real world, benefits represented by the amount of risk reduction to be achieved and the associated costs become important factors. Trade-offs are frequent and necessary. An appropriate goal in the decision-making process is for the residual risk to be as low as reasonably achievable. Paraphrasing the terms of the NRC definition of ALARA helps explain the process:

1) Reasonable efforts must be made to identify, evaluate, and eliminate or control hazards so that the risks deriving from those hazards are acceptable.

2) In the design and redesign processes for physical systems and work methods, risk levels for injuries and illnesses, and property and environmental damage must be as far below what would be achieved by applying current standards and guidelines as is economically practicable.

3) For items 1 and 2, decision makers must consider:

- purpose of the undertaking;
- state of the technology;
- costs of improvements in relation to benefits to be attained;
- whether the expenditures to reduce risk in a given situation could be applied elsewhere with greater benefit.

Spending an inordinate amount of money to reduce the risk only a little through costly engineering and redesign is inappropriate, particularly if that money could be better spent otherwise (such as for an exercise facility). The author leans a bit more toward ALARA as it is applied than ALARP as it is applied.

To the credit of the authors of ANSI/ASSE Z10-2012, Occupational Health and Safety Management Systems, the standard includes a provision requiring that risk assessments be conducted (ANSI/ASSE, 2012, p. 15).

If present terminology holds, ISO 45001, the forthcoming international standard for safety management systems, will include a provision requiring that risk assessments be made.

Making risk assessments is the core of ANSI/ASSE Z590.3-2011(R2016), Prevention Through Design: Guidelines for Addressing Occupational Hazards and Risks in Design and Redesign Processes. In that standard, the ALARP concept is embodied within its definition of acceptable risk:

> Acceptable risk: That risk for which the probability of an incident or exposure occurring and the severity of harm or damage that may result are as low as reasonably practicable (ALARP) in the setting being considered. (ANSI/ASSE, 2016, p. 12)

Similarities exist in the definition of ALARP in the CSB report as compared to the definition set forth in Z590.3, the latter of which follows.

> ALARP: That level of risk which can be further lowered only by an increase in resource expenditure that is disproportionate in relation to the resulting decrease in risk. (ANSI/ASSE, 2016)

That safety professionals will be expected to have sufficient knowledge and capability to make and give counsel on risk assessments has evolved over the past 40 years, albeit somewhat slowly.

Leading and Lagging Indicators

CSB makes much of leading indicators. Why? Both BP and Transocean had good OSHA-type incident rates. CSB makes the case, as have others, that having good incident rates does not necessarily indicate that an organization's risks related to serious injury and fatality potential are well controlled. CSB encourages organizations to have a set of metrics that relate to the performance level desired for the key processes, operating discipline, and layers of protection that relate to identified hazards and risks.

What CSB writes about leading indicators is refreshing. Consider the following two excerpts from Volume 3:

> •Indicators should measure the health of the company's safety management system and the *specific barriers in place to prevent or mitigate major accident hazards* [Emphasis added] (CSB, 2012). (CSB, 2016b, p. 152)

> •The selected indicators should be actionable in terms of the necessary actions to improve some specific aspect of safety performance. (CSB, 2016b, p. 153).

To prevent or mitigate major incident hazards, organization management must know the major hazards. To accomplish this, hazard analyses and risk assessments must be conducted. Results of the risk assessments would include determining the "actionable items" that are "to improve some specific aspect of safety performance" and to achieve risk levels as low as reasonably practicable.

Very little literature exists on leading indicators that direct safety practitioners to perform hazard analyses and risk assessments so that major incident potentials can be identified. The author identified one such document. Comments on this document appear in *On the Practice of Safety* (Manuele, 2013), in the chapter, "Measurement of Safety Performance."

> In 2006, the Health and Safety Executive in the U.K. published *Developing Process Safety Indicators: A Step-by-Step Guide for Chemical and Major Hazard Industries.* The approach taken in this guide with respect to leading and lagging indicators differs from any other approach discovered in this author's research.
>
> The guide introduces the idea of "dual assurance" in which leading and lagging indicators are set in a "structured and systematic way." Acting in concert, they serve as "system guardians providing dual assurance to confirm that the risk control system is operating as intended or providing a warning that problems are starting to develop."
>
> In this system, leading indicators are related directly to lagging indicators. And that makes good practical sense. Since the purpose of an operational risk management system is to reduce in so far as is practicable the occurrence of what are called lagging indicators, then energies expended on managing leading indicators that relate directly to lagging indicators is properly directed. (p. 550)

Although the guide applies to the chemical and major hazard industries, the author recommends it to all safety practitioners involved in performance measurement, and leading and lagging indicators because its thought processes are worthy of consideration.

Incident Barriers

That the CSB report pleads for appropriate barriers to be in place to prevent major incidents is significant and provides safety practitioners an opportunity for reflection. This subject needs greater recognition in the safety practitioner community.

A previously cited excerpt from Volume 3 is repeated here because it has bearing on the discussion of barriers that follows.

> Indicators should measure the health of the company's safety management system and the *specific barriers in place to prevent or mitigate major accident hazards* [Emphasis added] (CSB, 2012). (CSB, 2016b, p. 152)

Hollnagel's (2004) comments on barriers are significant in relation to having "specific barriers in place to prevent or mitigate major accident hazards."

> A barrier is, generally speaking, an obstacle, an obstruction, or a hindrance that may either: 1) prevent an event from taking place, or 2) thwart or lessen the impact of the consequences if it happens nonetheless. In the former case the purpose of the barrier is to make it impossible for a specific action or event to occur. In the latter case the barrier serves, for instance, to slow down uncontrolled releases of matter and energy, to limit the reach of the consequences, or to weaken them in other ways.
>
> Barriers are important for the understanding and prevention of accidents in two different, but related, ways. Firstly, the very fact that an accident has taken place usually means that one or more barriers have failed—either because they did not serve their purpose adequately or because they were missing or dysfunctional.
>
> The search for barriers that have failed must therefore be an important part of accident analysis. Secondly, once the aetiology of an accident has been determined and a satisfactory explanation has been found, barriers can be used to prevent the same or similar accidents from taking place in the future. In order to facilitate this, the consideration of barrier functions must be a natural part of system design. (p. 68) *[Note: As used here, aetiology is the study of causation.]*

Hollnagel (2004, p. 79) references Haddon's concept and strategies concerning the avoidance of unwanted energy releases.

William Haddon, the first director of the National Highway Safety Bureau, proposed that his energy release theory was applicable in preventing incidents and reducing the severity of injury or damage if an incident occurred. Its concept is that unwanted transfers of energy can be harmful (and wasteful) and that an organization should take a systematic approach in the design and operating processes to limit their possibility. Part of this approach consists of providing physical or procedural barriers to prevent contact by persons or property and to direct an energy flow into wanted channels. Haddon's barrier concepts are soundly based.

Haddon (1970) states that "the concern here is the reduction of damage produced by energy transfer." But he also says that "the type of categorization here is similar to those useful for dealing systematically with other environmental problems and their ecology." Excerpts from Haddon (1970) follow. Note that all of the strategies relate to facility design or work methods design.

> A major class of ecologic phenomena involves the transfer of energy in such ways and amounts, and at such rapid rates, that inanimate or animate structures are damaged.
>
> Several strategies, in one mix or another, are available for reducing the human and economic losses that make this class of phenomena of social concern. In their logical sequence, they are as follows:
> - Prevent the marshaling of the form of energy.
> - Reduce the amount of energy marshaled.
> - Prevent the release of the energy.
> - Modify the rate or spatial distribution of release of the energy from its source.
> - Separate, in space or time, the energy being released from that which is susceptible to harm or damage.

- Separate, by interposing a material barrier, the energy released from that which is susceptible to harm or damage.
- Modify appropriately the contact surface, subsurface, or basic structure, as in eliminating, rounding, and softening corners, edges, and points with which people can, and therefore sooner or later do, come in contact.
- Strengthen the structure, living or nonliving, that might otherwise be damaged by the energy transfer.
- Move rapidly in detection and evaluation of damage that has occurred or is occurring, and counter its continuation or extension.
- After the emergency period following the damaging energy exchange, stabilize the process. (p. 229)

All hazards are not addressed by the unwanted energy release concept. Examples are the potential for asphyxiation from entering a confined space filled with gas, or inhalation of asbestos fibers. But all hazards are encompassed within a goal that is to avoid *both* unwanted energy releases *and* exposures to hazardous environments.

Many improvements made in the interior and exterior design of automobiles to reduce the occurrence of incidents and their potential severity relate to Haddon's principles.

Hollnagel (2004) also refers to management oversight and risk tree (MORT) as an indication of a treatment and a resource with respect to barriers.

> MORT proposed a distinction between several different types of barriers. These were: 1) physical barriers; 2) equipment design; 3) warning devices; 4) procedures/work processes; 5) knowledge and skills; and 6) supervision. Finally, the MORT barrier analysis discussed how barriers might be unable to achieve their purpose, either because they failed as such or for other reasons. (p. 80)

ANSI/ASSE Z10-2012, Occupational Health and Safety Management Systems, is an additional and valuable resource with respect to barriers. In its planning section, safety and health issues are defined as "hazards, risks and management system deficiencies" (ANSI/ASSE, 2012, p. 9).

Barriers are defined here in the widest possible scope. Hollnagel's (2004) comments on MORT are a good reference for what barriers may include. Barriers include all aspects of operations that relate to hazards and the risks that derive from them, and the relative management systems that should be in place to achieve acceptable (i.e., as low as reasonably practicable) risk levels.

If appropriate barriers and controls exist and the management systems pertaining to them have no deficiencies, then damaging incidents are less likely to occur. Having appropriate and well-managed barriers and controls in place is critical for every aspect of operational risk management.

Management of Change

Following is an excerpt from Volume 3 (CSB, 2016b).

> Experience shows that changes in the operating environment, systems, procedures, equipment, organization, and management personnel and practices represent some of the biggest challenges to effectively managing major hazard risks. Poorly managed change frequently results in serious failures, many of which are precursors to major accidents (or higher costs as well). A vital component of change management is an assessment of how those technical changes may influence human performance. (p. 102)

OSH practitioners should closely examine the preceding excerpt. It is loaded. Although the comments are within a report on an offshore disaster, they pertain to a broad range of operations.

Studies of incident investigation reports show that "poorly managed change frequently results in serious failures, many of which are precursors to major accidents (or higher costs as well)" is too often the case. CSB (2016b) says "a missed opportunity" was an important factor in the Macondo event (p. 103).

If safety practitioners believe their responsibilities include a focus on serious injury and fatality prevention, it is to their advantage to review for adequacy the management of change practices in the organizations they advise. If effective management of change processes are in place, fewer incidents resulting in serious injuries or fatalities would occur.

Recognizing Practical Limitations

In its report, CSB (2016b) boldly suggests that designers of systems and writers of standard operating procedures may not be able to achieve perfection.

> The operator cannot write a drilling program that foresees all circumstances and covers every detail for the drilling contractor to follow. Therefore, the operator and drilling contractor must actively work to bridge the gap between work-as-imagined in the drilling program as defined by well designers, managers, or even regulatory authorities and work-as-done by the well operations crew. (p. 84)

> Gaps between work-as-imagined . . . and work-as-done . . . must be continually identified, managed and minimized by building a resilient process that can sustain desirable operations during both expected and unexpected conditions. (p. 84)

While the preceding excerpts refer to drilling and contractors, their premise applies to all types of operations. To expect designers and writers of standard operating procedures to achieve perfection and be able to envision all possible hazards denies their humanity. Gaps typically exist between work-as-imagined and work-as-done.

Why is this significant? Safety practitioners should realize that variation from what is prescribed is a norm. When an incident occurs and it is learned that an employee did not follow the standard operating procedure, a five-why process should be applied to determine the reason for that person's actions. It may be that what was done seemed logical to the employee and the supervisor as they made revisions in the work process.

Incident Investigation

The following excerpt from Volume 3 is titled "Expanding Beyond Immediate Causes and Implementing Change." Safety professionals should take note of the sound concepts it presents.

> The broadest learning impact can be achieved when investigations extend beyond the immediate technical causes of an incident. Addressing deficient safety management systems and inadequate organizational practices can result in findings that go beyond the immediate chain events that preceded any one incident. As examples in this chapter show, while the immediate causes of a well control incident might vary, the safety management systems and organizational findings can be similar. . . . There is the danger of concentrating on the exact mechanism of the previous incident rather than identifying broad lessons. (p. 127)

Regulatory Recommendations

Volume 4 is titled "Regulatory Oversight of U.S. Offshore Oil and Gas Operations: A Call for More Robust and Proactive Requirements." The document fulfills CSB's responsibilities as described in the following excerpt.

> The CSB's preventive mission as a federal agency is to reduce chemical hazards as broadly as possible (e.g., through recommendations that will effect national preventive changes). The CSB, therefore, focuses its recommendation efforts on changing national legislation, regulation, voluntary consensus standards and industry recommended practices. As a result of an investigation or study, the CSB may issue "proposed rules or orders" to regulators such as the EPA Administrator and the Secretary of Labor "to prevent or minimize the consequences of any release of substances that may cause death, injury or other serious adverse effects on human health or substantial property damage as the result of an accidental release" [42 U.S.C. 7412(r)(6)(c)(ii)]. (CSB, 2016c, p. 12)

CSB (2016a, pp. 12-23) summarizes "Key Investigative Findings and Conclusions," including those pertaining to regulatory attributes. The following excerpts from Volume 4 reflect what CSB (2016c) continues to stress.

> For example, key findings in Volumes 3 and 4 of the Macondo Report show that the U.S. offshore regulator lacks effective use of key process safety indicators and guidance ad-

dressing corporate boards of directors and human factors focused on major accident prevention. The CSB report analysis shows that addressing these significant gaps could help reduce the risk of similar incidents. (p. 12)

Under the heading "Continual Risk Reduction to Levels As Low As Reasonably Practicable (ALARP)," CSB (2016c) says:

> The intention of a goal-based, risk-reduction regulatory framework is to eliminate or sufficiently minimize the risks in an operation. Although risk can never be completely eliminated, any such framework must continually strive toward this goal. With major accident hazards, the key question becomes: Is there anything more that can be done to reduce the risk? ALARP is a standard familiar to industry in other global offshore regimes, and even in other high-hazard industries in the U.S. (p. 14)

As part of the agency's investigative approach, CSB may examine the strengths and weaknesses of regulations that other countries have adopted. CSB reviewed regulatory requirements in the U.K., Australia and Norway, and found that U.S. requirements have gaps that CSB hopes will be eliminated as they are revised (p. 17).

Conclusion

CSB's report on the *Deepwater Horizon* incident is fascinating for the positions it takes. The writers of the report indirectly advance the state of the art in safety management. Safety practitioners will benefit by asking what they can learn from the executive summary and four volumes that make up the CSB report. ■

References

ANSI/ASSE. (2012). Occupational health and safety management systems (ANSI/ASSE Z10-2012). Retrieved from www.asse.org/ansiaihaasse-z10-2012-occupational-health-safety-management-systems/ansi/aiha/asse-z10-2012-occupational-health-and-safety-management-systems

ANSI/ASSE. (2016). Prevention through design: Guidelines for addressing occupational hazards and risks in design and redesign processes [ANSI/ASSE Z590.3-2011(R2016)]. Retrieved from www.asse.org/ansi/asse-z590-3-2011-r2016-

Center for Chemical Process Safety (CCPS). (2007). *Guidelines for risk based process safety*. Hoboken, NJ: Wiley.

CSB. (2012, July 23-24). CSB public hearing: Safety performance indicators, Houston, TX, p. 183. Retrieved from www.csb.gov/assets/1/19/CSB_20Public_20Hearing.pdf

CSB. (2014a, June 5). Investigation report volume 1: Explosion and fire at the Macondo well (Report No. 2010-10-I-OS). Retrieved from www.csb.gov/macondo-blowout-and-explosion

CSB. (2014b, June 5). Investigation report volume 2: Explosion and fire at the Macondo well (Report No. 2010-10-I-OS). Retrieved from www.csb.gov/macondo-blowout-and-explosion

CSB. (2016a, April 12). Investigation report executive summary: Drilling rig explosion and fire at the Macondo well (Report No. 2010-10-I-OS). Retrieved from www.csb.gov/assets/1/7/20160412_Macondo_Full_Exec_Summary.pdf

CSB. (2016b, April 12). Investigation report volume 3: Drilling rig explosion and fire at the Macondo well (Report No. 2010-10-I-OS). Retrieved from www.csb.gov/assets/1/7/Macondo_Vol3_Final_staff_report.pdf

CSB. (2016c, April 12). Investigation report volume 4: Drilling rig explosion and fire at the Macondo well (Report No. 2010-10-I-OS). Retrieved from www.csb.gov/assets/1/7/Macondo_Vol4_Final_staff_report.pdf

Haddon, W. (1970, Dec.). On the escape of tigers: An ecologic note. *American Journal of Public Health, 60*(12).

Health and Safety Executive (HSE). (2006). *Developing process safety indicators: A step-by-step guide for chemical and major hazard industries.* Retrieved from www.hse.gov.uk/pubns/books/hsg254.htm

Hollnagel, E. (2004). *Barriers and accident prevention.* Burlington, VT: Ashgate.

Hopkins, A. Why "safety cultures" don't work. Sydney, Australia: Future Media Training Resources. Retrieved from www.futuremedia.com.au/docs/Why_Safety_Cultures_Don't_Work.pdf

Knox, N.W. & Eicher, R.W. (1992, Feb. 1). MORT user's manual: For use with the management oversight and risk tree analytical logic diagram (Technical report No. DOE/SSDC-76-45/4-Rev.3; SSDC—4-Rev.3). Washington, DC: U.S. Department of Energy. Retrieved from www.osti.gov/scitech/biblio/5254810

Manuele, F.A. (2013). *On the practice of safety* (4th ed.). Hoboken, NJ: Wiley.

Manuele, F.A. (2014). *Advanced safety management: Focusing on Z10 and serious injury prevention* (2nd ed.). Hoboken, NJ: Wiley.

Noordwijk Risk Initiative Foundation. (2009, Dec. 20). NRI MORT user's manual (2nd ed.) [No. NRI-1 (2009)]. Delft, The Netherlands: Author. Retrieved from www.nri.eu.com/NRI1.pdf

Sutcliffe, K.M. (2011, Oct. 17). Expert report for the U.S. District Court for the Eastern District of Louisiana (MDL No. 2179), Section: J, re. Oil spill by the oil rig *Deepwater Horizon* in the Gulf of Mexico, on April 20, 2010 (p. 92).

U.S. Nuclear Regulatory Commission (NRC). (2017). ALARA. Retrieved from www.nrc.gov/reading-rm/basic-ref/glossary/alara.html

Wilkinson, P. (2016, April). Culture: Values and practices—Can you have one without the other? Retrieved from www.csb.gov/assets/1/19/Wilkinson_-_Values_and_Practices_FINAL.pdf